1-6-75

Residential Fuel Policy and the Environment

Residential Fuel Policy and the Environment

Alan S. Cohen

Gideon Fishelson

John L. Gardner

Ballinger Publishing Company • Cambridge, Mass.
A Subsidiary of J.B. Lippincott Company

International Standard Book Number: 0-88410-324-2

Library of Congress Catalog Card Number: 74-7213

Printed in the United States of America

Library of Congress Cataloging in Publication Data
Cohen, Alan S 1944–
Residential fuel policy and the environment.
Includes bibliographies.
1. Energy policy—United States. 2. Dwellings—Heating and ventilation—Environmental aspects—United States. 3. Fuel—Prices—United States. I. Fishelson, Gideon, joint author. II. Gardner, John L., joint author. III. Title.
HD9502.U52C64 333.8'2'0973 74-7213
ISBN 0-88410-324-2

Contents

List of Figures

List of Tables

Foreword

Energy problems following closely after the rise in concern for the environment have caused an intense awareness of the costs involved in achieving environmental aims. There is a need to consider environmental measures anew, to try to judge how far to press various types of environmental improvement in view of the sacrifices entailed.

A chief impediment to serious evaluation of environmental measures in the past has been disciplinary fragmentation. Environmental effects depend on physical science relationships whose modeling requires engineering. Estimating the costs and benefits requires economics. This is particularly true for benefits which involve complex valuation problems. An inter-disciplinary team approach is needed to achieve a comprehensive perspective for such an evaluation. The study presented here is an engineering–economic investigation which has met this need by developing a compatible set of economic and engineering models.

The study is concerned with an air pollution problem of traditional significance, namely, air quality deterioration from home heating. Policy conclusions are reached indicating the need for household fuel measures in the face of energy problems. As a contribution that is in many ways more important, methods are developed for estimating the benefits and costs of environmental policies. The methods have wide applicability beyond the immediate subject matter focused upon here.

The underlying research is from a project entitled "Environmental Pollutants and the Urban Economy," funded under the National Science Foundation RANN program. The grant was made jointly to the University of Chicago and Argonne National Laboratory.

G.S. Tolley
Kevin Croke

Acknowledgments

The authors wish to acknowledge the following organizations for their assistance and cooperation in acquiring the data needed for this analysis: Chicago Department of Environmental Control, Commonwealth Edison Company, Federal Environmental Protection Agency–Region V, Illinois Environmental Protection Agency, and Peoples Gas Light and Coke Company. We are grateful to Thomas Baldwin, Lyndon Babcock, and Niren Nagda for their supporting studies used in this research, and to Mary Snider and James Cavallo for their computer programming assistance. Finally, we would like to thank the National Science Foundation for funding the research underlying this book.

Residential Fuel Policy and the Environment

Chapter One

The Setting

INTRODUCTION

The relationship between energy consumption and environmental quality is a national concern. Sweeping legislation passed in 1967 and 1970 set up mechanisms for controlling air pollution, including the environmental effects of energy generation and use. Federal, state, and local environmental programs mandated by this legislation have affected fuel and energy production and consumption patterns, and in some cases have contributed to shortages in natural gas, gasoline, certain types of coal, and electrical energy. The conflicting demands for energy and for better air quality necessitate trade-offs between the availability and use of fuel in our economy and the pace and vigor of environmental protection programs.

The evaluation of environmental programs should include a framework that will identify the costs and benefits of alternative policies relating to energy use for each sector of the economy. The procedures originally employed to devise energy-related environmental programs did not, at least explicity, include such benefit–cost evaluations. The objective of this study is to develop a benefit–cost evaluation framework for one major policy set, namely, residential fuel policies.

This study develops an energy–environmental policy evaluation model containing several components: (1) business and household responses to policy regulations for controlling emissions, (2) costs associated with these responses, (3) impacts on air quality with a high level of geographical detail, and (4) the value of beneficial effects on human health, materials, and the aesthetic quality of the environment. The model is applied in this study to environmental policies affecting residential space heating in Chicago. This specific policy evaluation is an illustration of the model's usefulness for conducting an integrated engineering–economic analysis of environmental policies in an urban area.

The approach to policy evaluation advocated in this study is unique in giving attention to the total dollar value of air quality benefits achieved relative to aggregate costs, and to the distribution of both benefits and costs among locations and among identifiable groups within the region affected by the policy. This mode of analysis is different from the method of evaluating policies currently in use by national and state environmental control agencies. The latter requires attention only to relative costs of compliance, among all policies that improve environmental quality everywhere in a region to below given standards. Figure 1–1 outlines the model and indicates where in the book each component is discussed in detail. (The letters in the boxes indicate appendixes in which technical details, empirical results, and extensions for each chapter are provided.)

In Chapter 2, the social costs associated with each residential fuel policy are identified, and estimates of these costs are presented. The impacts of these costs on homeowners, apartment owners, tenants, and fuel distributors are analyzed through 1990. The effects of changes in fuel prices are also explored. Finally, the sensitivity of costs to the underlying assumptions is analyzed with attention being given to different compliance schedules, conversion costs, and abandonments.

In Chapter 3, the air quality impacts of these alternative policies are analyzed by using an air quality dispersion model. The effects of residential space heating fuel consumption on air quality are identified, and a brief description of the methods of modeling the relationship between emissions and air quality is given. The air quality improvements due to each of the residential policies and industrial, utility, and commercial regulations are presented in a series of air quality maps of Chicago. These estimates are compared to federal standards, and the confidence of meeting the standards is discussed. The regulations used in this study are presented in Appendix A.

In Chapter 4, the social benefits of the alternative residential fuel policies are estimated. First, the effects of air quality improvement on death rates, costs of health care, and costs of maintaining materials are estimated. Second, dollar values of these effects are estimated. An alternative computation of benefits based on the effects of air pollution on property values is presented.

In Chapters 5 and 6, the cost estimates presented in Chapter 2 and the benefit estimates presented in Chapter 4 are compared. The best residential fuel policy is identified to be a ban on the use of coal by 1975. Chapter 5 discusses the economic impacts of this policy, with special attention to the distribution of net benefits by income and race, and the distribution of costs by economic sector. In the course of developing the model and considering its application in the context of environmental policies in Chicago, several issues of wider significance are raised. These include impacts of rising fuel prices

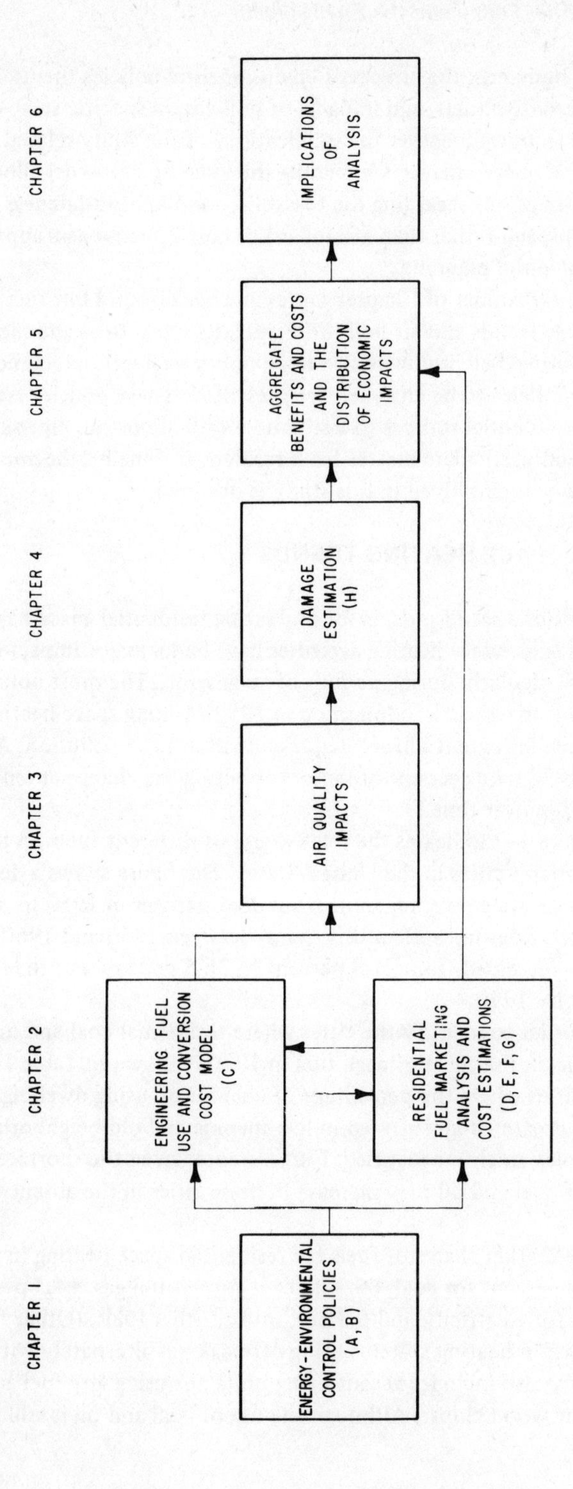

Figure 1–1. Schematic of the Energy–Environmental Policy Analysis.

on local energy demands, the effects of environmental policies themselves on prices of alternative fuels, and impacts of policies on specific socio-economic classes. Chapter 6 draws together the implications of the study related to national and local energy markets including the issue of clean fuel allocation, and presents conclusions regarding the feasability and appropriateness of a benefit–cost approach rather than a standard or cost-effectiveness approach to local environmental planning.

The remainder of Chapter 1 presents background information on residential fuel usage trends and air pollution regulatory practices and identifies the need for improved environmental–energy policy evaluation techniques. The residential fuel policies to be analyzed are presented. These policies range from no controls on residential sources to a situation with almost no emissions of sulfur dioxide and particulate matter from residences. Finally, the approach to benefit–cost analysis employed in this study is outlined.

RESIDENTIAL SPACE HEATING TRENDS

This study develops a set of models for evaluating residential space heating fuel policies. Historically, space heating activities have had a major impact on air quality levels, particularly during periods of stagnation. The most notable example of this is the great London smog in 1952. Among space heating fuels, coal, and to a smaller extent oil, are major contributors to pollution. A study of fuel policies should take account of factors affecting the change among fuels and changes in fuel use over time.

Figure 1–2 indicates the relative use of different fuels on the average in twenty-four major cities in the United States. The figure shows a decline in the percentage of coal-using dwelling units from 38.5 percent in 1950 to 3.4 percent in 1970. Oil usage does not reflect this trend; between 1950 and 1960 its share in space heating increased from 21.8 percent to 36.5 percent and then declined to 28.4 percent by 1970.

In order to identify the cities where residential coal and oil policies might be beneficial, the share of each fuel in 1970 is shown in Table 1–1. Even in those cities where the percentage of coal- or oil-using dwelling units is small, their use is often concentrated in low-income and old neighborhoods in which a fuel policy might be justified. Furthermore, given the shortages of clean fuels, the use of coal and oil may increase in these cities in the absence of any fuel use policy.

The market shares of fuels for residential space heating in Chicago in 1972 were 12.0 percent for coal, 66.3 percent for natural gas, 14.3 percent for oil, 1.1 percent for electricity and 6.3 percent for other fuels. (Other fuels are mainly combination heating systems using off-peak gas alternately with coal or oil. This category also includes vacant living units not using any fuel and units using propane or wood chips.) Although the use of coal and oil is still significant,

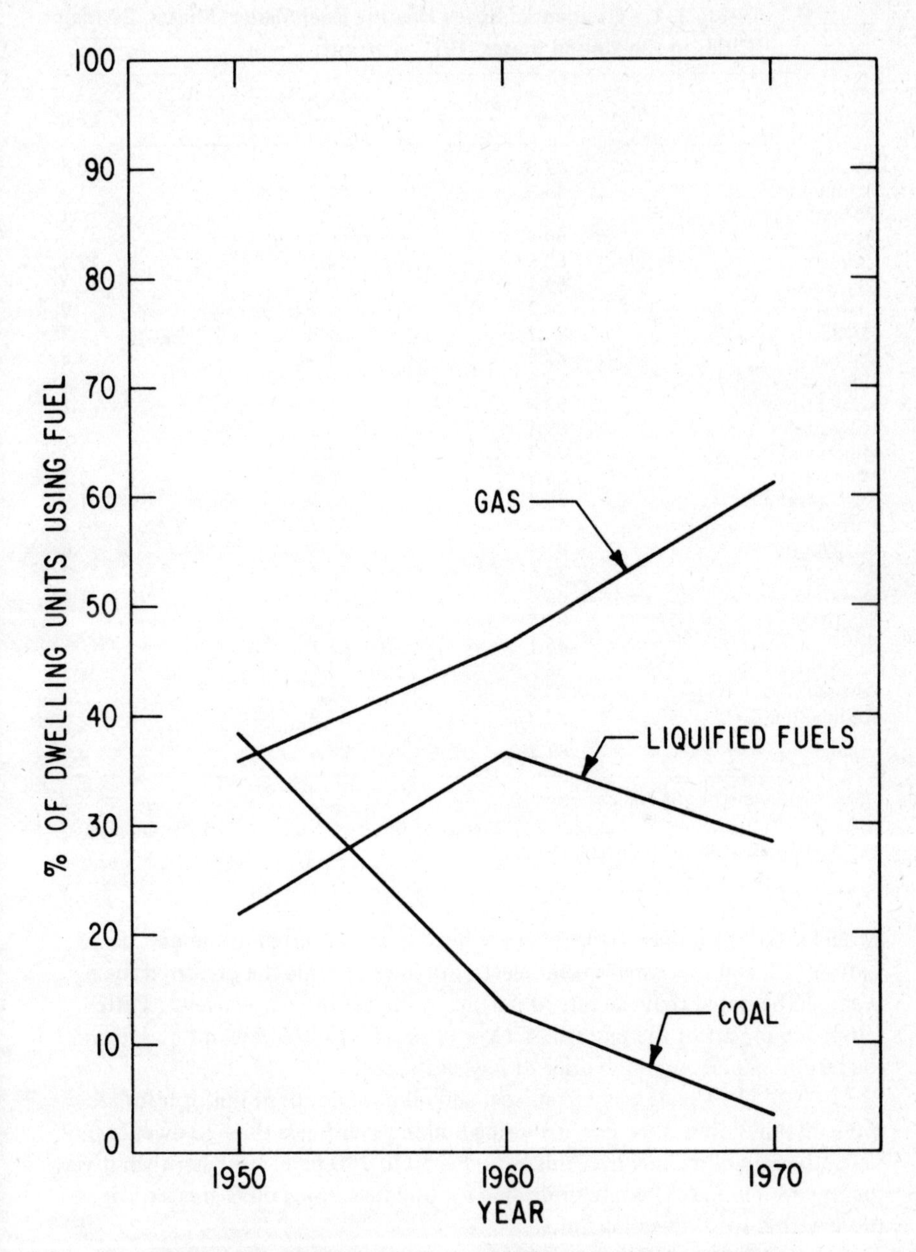

Figure 1-2. Residential Space Heating Fuel Market Shares in Twenty-Four Major Cities.

Table 1-1. Residential Space Heating Fuel Market Shares, 24 Major Cities in the United States, 1970 (Percent)

City	*Gas*	*Liquified Fuels*	*Coal*
Atlanta	87.0	1.0	0.5
Baltimore	44.3	49.2	1.4
Boston	30.3	63.7	.4
Buffalo	90.4	6.6	.8
Chicago	62.5	14.8	17.4
Cincinnati	89.1	2.0	2.3
Cleveland	94.2	.9	.9
Dallas	85.7	*	.2
Detroit	87.4	4.5	4.6
Houston	84.8	*	.2
Kansas City	93.4	1.2	.2
Los Angeles	89.1	.2	*
Milwaukee	72.2	21.1	3.2
Minneapolis	86.9	6.5	2.7
New York	29.7	63.5	1.9
Philadelphia	57.5	35.2	3.2
Pittsburgh	95.1	.8	.6
Portland	24.6	56.9	.2
St. Louis	86.4	7.9	1.6
San Diego	87.8	.3	*
San Francisco	86.1	1.7	*
Scranton	38.9	23.4	32.9
Seattle	23.6	51.8	.3
Washington, D.C.	52.3	34.9	3.8
Total	61.1	28.4	3.4

*Less than one-tenth of one percent.
Source: U.S. Bureau of the Census, *1970 Census of Housing*, Vol. 1, Part 1, Washington: U.S. Government Printing Office, 1972.

a trend away from these fuels for space heating has occurred in the past decade. Natural gas, and to a small extent electricity have become the preferred fuels. It should be noted that the rate of decline in the use of coal accelerated after 1968, due in part ot the passage of a low (1 percent) sulfur law in Chicago and the resulting increase in the price of low sulfur coal.

The trends away from coal and oil have not been uniform for all residential building sizes. For coal-using buildings with less than 20 dwelling units, the rate of decline over this period is 30 to 100 percent higher than it was for larger buildings. The rate of decline for buildings using oil is greatest for those with 8 to 19 dwelling units.

A survey of coal users in Chicago substantiates the hypothesis that the accelerated rate of conversion from coal after 1968 can be attributed, in part, to the Chicago low sulfur law and the associated price increase of coal. Evidence of this is shown in Figure 1-3, which indicates that the two most fre-

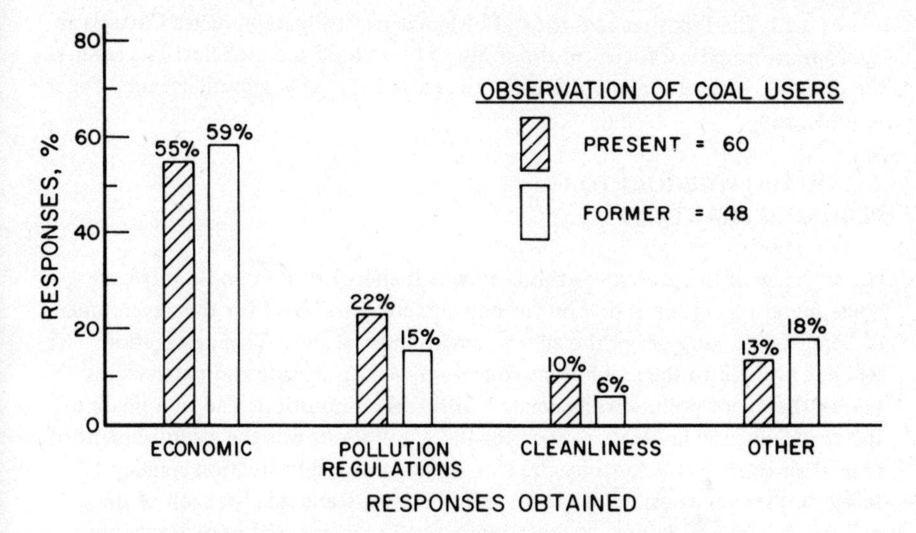

Figure 1–3. Reasons Given for Converting from Coal.

quently stated reasons for making conversions were economic and pollution regulations. For more details on the coal users' survey, see Appendix B.

Reflecting this decline in coal usage for residential space heating, the number of Chicago area coal dealers has been steadily declining. The number of coal distributors in Chicago has dropped from 253 in 1950 to 85 in 1967. A similar, though not as dramatic, decline has occurred in fuel oil distributorships: 215 distributors existed in 1958, but only 157 in 1967.

Several factors have contributed to the establishment of these trends. Among these are:

1. Regulations placed on the wellhead price of natural gas in 1962 artificially kept the price of this fuel down, while regulatory practices have given priority to residential uses of gas.
2. Competition among oil jobbers and among the coal merchants made advertisement by any one of them for a particular fuel non-profitable. On the other hand, advertising was very effective for both electricity and natural gas. This fuel market structure in which oil and coal are highly competitive, while electricity and natural gas are supplied by regulated monopolies, explains part of the past conversions from oil and coal to gas even when in certain situations a direct cost calculation would not justify such action.
3. Natural gas and electricity are considered to be convenient fuels, requiring little maintenance and janitorial labor. A conversion to these fuels therefore could result in savings in labor.

The fact that coal use in Chicago is relatively high makes Chicago an appropriate empirical focus for the study of household fuel policies. The results are directly relevant for other cities to the extent that coal and oil are used for space heating.

AIR QUALITY REGULATORY PLANNING PRACTICES

An overview of the present methods by which energy-related pollutants have come under regulation is helpful for understanding the need for the development of benefit-cost analyses in the energy-environmental field. This description will be confined to the regulations concerning sulfur dioxide and particulates, two of the major pollutants associated with fuel combustion. The first phase in the environmental control program for these pollutants was the establishment of federal air quality standards by the U.S. Environmental Protection Agency.[1] In 1970, the agency established two sets of annual standards for each of these pollutants. One set, known as the primary standards, is based upon epidemiological studies that set a threshold concentration in terms of micrograms per cubic meter ($\mu g/m^3$) above which deleterious health effects had been evidenced. The primary standard for particulates was set at 75 $\mu g/m^3$, annual geometric mean, of measured air quality; the primary standard for sulfur dioxide is 80 $\mu g/m^3$, annual arithmetic mean.

The secondary standards, based upon studies that determined the threshold above which *any* deleterious effects of pollutant concentrations would occur, are more stringent. Such effects include damage to property or vegetation. The secondary standard for particulates was set at 60 $\mu g/m^3$, annual geometric mean. The secondary standard for sulfur dioxide (60 $\mu g/m^3$, annual arithmetic mean) was recently abolished. This was done, in part, to ease the demands for scarce low sulfur fuels.

Once the federal air quality standards for these pollutants were established, each state was required to submit an implementation plan that included regulations already promulgated by the state for controlling particulate and sulfur dioxide emissions. In order for a state's implementation plan to be approved by the federal government, the state had to demonstrate that its emission control regulations would guarantee the attainment of the federal air quality standards by 1975. Several forms of emission control regulations exist for fuel combustion activities. Among these are a restriction on the ash or sulfur contents of the fuels, bans on the use of certain fuels, and restrictions upon the pounds of pollutants that could be emitted per million British thermal units (Btu) of energy generated by a fuel combustion process.

Each state, in order to select the appropriate regulations for utility, industrial, commercial and residential fuel combustion sources, was required to make an inventory of the pollution-producing sources within its boundaries.

Given the emission characteristics of these sources, obtained from this survey, the control agency would determine how stringent the emission-reduction program would have to be in order to meet federal air quality standards by 1975. The minimum amount of emission reduction consistent with the attainment of the air quality standards and the state's environment goals was then selected for the state's implementation plan. In lieu of any cost information, this was assumed to be the minimum cost solution.

In estimating the reductions to be realized from emission control regulations, it was assumed that each fuel combustion emission source would reduce emissions either by the installation of pollution control equipment (which would remove particulate matter or sulfur dioxide from the flue gases) or by switching from a fuel that has high emission characteristics to a cleaner fuel. In the case of the residential space heating sector, no control devices presently exist for units within the range of sizes of these sources. Consequently, the only way residential space heating pollution sources can comply with air quality control regulations is to switch from high-emission to low-emission fuels. For the control of particulate emissions, this action would imply a switch from the use of high ash coal to either low ash coal, oil, natural gas, or electricity. In the case of sulfur dioxide, this would imply a conversion from the use of high sulfur coal or oil to low sulfur coal or oil, natural gas, or electricity.

As a result of both state implementation plans and the independent inititative of many municipal governments, a number of stringent environmental control regulations were passed regarding the use of fuels for residential space heating. Two characteristics of these ordinances should be noted. The first is that many of them were passed between 1965 and 1972. Secondly, many of these ordinances explicitly or implicitly banned the use of high sulfur fuels, which caused a rapid shift in the residential demand for low sulfur energy sources.

If the federal air quality program is to be re-evaluated in terms of its effects on the supply of fuel for residential, industrial, and utility use, three characteristics of the environmental regulatory process described above become of critical importance. The *first* of these is related to the way in which the required emission reductions for any air quality control region are made. To comply with the requirements of The Air Quality Act,[2] each state made a survey of emission sources and calculated the required emission reductions based upon this survey. These projections, however, did not account for the fact that patterns of energy generation and use are constantly changing as a result of the dynamics of the energy market. Technological changes, scarcity of clean fuels, changes in the price structure, and government regulations affecting the energy industry may cause significant shifts in energy utilization over the planning period used in the implementation process. If these changes are not considered when the required reductions of emissions are calculated, then inappropriate regulations may be passed.

The *second* critical problem posed by present environmental regulatory planning procedures is that the effects in terms of air quality and the costs of changing the timing of the attainment of air quality standards are not explicity identified. In particular, in the control of emissions in the residential space heating sector, fuel-switching entails the premature conversion or retirement of heating equipment associated with the use of high emission fuels. Both the cost and effectiveness of a residential emission control program are quite sensitive to the length of time permitted for compliance. Meeting environmental control requirements may involve the investment of capital whose amortization may take place over a fifteen-year period or longer. Thus, in evaluating the effects of environmental–energy–related control programs, it is not sufficient to consider only the environmental control costs and air quality improvements through 1975. A longer planning horizon is required.

The *third* major problem associated with present environmental regulatory planning process is the procedure by which the levels of acceptable air quality were established. The acceptance of threshold effects as the basis for allowable air quality standards is inconsistent with potential trade-offs between the value of air quality improvements and the cost of achieving these improvements. In order to assess whether certain forms of energy with higher emission characteristics should be allowed to be used during periods of tight energy supply, it is necessary to quantify the dollar value of incremental changes in air quality levels. The benefit–cost framework for evaluating environmental regulations to the residential sector, as reported in this study, will specifically address these problems.

POLICIES TO BE EVALUATED IN THIS STUDY

The number of potential residential policies aimed at improving air quality is actually infinite; policies range from no control to those requiring the use of a clean fuel for space heating by all residences. One can examine a continuous policy scheme starting with banning coal use in single family homes, then in duplexes, three-flats, and so on. A similar approach can be worked out for oil. Another avenue is to assign different compliance dates to different customers. In this study the problem is made manageable by examining a few discrete cases that encompass the range.

The choice of policies considered grew out of the history of residential fuel regulations. In the public hearings on the Illinois implementation plan in December, 1971,[3] evidence was presented that indicated the federal secondary particulate standard could not be met in certain areas of the city without a ban on the use of coal for residential space heating. Although total emissions by weight are not overly significant, the characteristics of residential space heating pollution sources—e.g., emission of pollutants relatively close to ground level—magnify their impact on air quality. Therefore, the Illinois Pollution Control Board promulgated regulations that in effect banned the use of coal for residential space heating in the Chicago metropolitan area.

Subsequent to the submission of the implementation plan, the Coal Merchants' Association filed suit in the Cook County Circuit Court and obtained an injunction against the imposition of a coal ban on the basis that such a ban would generate undue hardship to the coal-distributing industry within the Chicago area. Since the state was temporarily halted from the passage of what was considered to be a necessary regulation to attain the federal air quality standards, the U.S. Environmental Protection Agency did not approve the Illinois implementation plan. Because the state plan was not approved, the administrator of the U.S. Environmental Protection Agency, under the provisions of the Clean Air Act of 1967 and its amendment in 1970, took the first steps in the promulgation of regulations that would impose a ban on use of coal for residential space heating in the city of Chicago. The federal plan was restricted to the city of Chicago on the basis that major environmental problems related to the use of coal for space heating did not exist outside the city of Chicago. From the *1970 Census of Housing*, fuel market shares in the Chicago metropolitan area, *excluding* the city of Chicago, were calculated.[4] These were 83.0 percent for gas, 12.4 percent for oil, 3.3 percent for electricity, and only 1.3 percent for coal in 1970, which indicates that clean fuels were predominantly used outside the city limits.

Using this history of residential space heating control as a guide, four alternative policy options were defined for the benefit–cost evaluations:

1. *No restrictions on the burning of fuels for residential space heating.* This policy was included as a bench mark for comparison with the effects of the other residential fuel policies.
2. *A ban on the use of fuels having more than 1 percent sulfur content by weight*. This regulation was imposed by the city of Chicago in 1968. A gradual reduction in the allowable sulfur content of fuels was required by this ordinance between 1968 and 1972. By 1972, no fuel above the 1 percent sulfur limit could be burned.
3. *A ban on the use of coal for residential space heating and a 1 percent sulfur restriction for oil.* This policy essentially reflected the requirements of the Illinois implementation plan and the proposed federal regulation for Chicago. Two compliance dates were considered, 1975 and 1977.
4. *A ban on the use of coal and oil for residential space heating.* This constituted a more stringent regulation than that required by the implementation plan. The banning of both coal and oil use in the residential market reduces the emission of particulates and sulfur dioxide from this sector to practically zero. Again, the policy was evaluated for compliance dates of 1975 and 1977.

ENVIRONMENTAL BENEFIT-COST FRAMEWORK

A benefit-cost analysis to evaluate the effects of environmental regulations on the residential energy market entails the development of a method for estimating social costs of air pollution control. Such analyses must include the costs of changes in fuel use patterns that can be attributed to air pollution regulations, as well as the outright retirement of space heating equipment or capital associated with the combustion and distribution of high emission fuels. Additional costs may be incurred if residential buildings are abandoned prematurely due to environmental regulations. Furthermore, environmental benefit-cost analysis requires a method to evaluate changes in morbidity and mortality rates and damage to materials resulting from changes in air quality levels.

Postulating that such changes in the social costs and benefits can be aggregated over all individuals, then benefit and cost functions, as illustrated in Figure 1-4, can be constructed. The height of the benefit curve at A, for example, measures the total reduction in social costs of health care, materials maintenance, and other damages when pollution restrictions are imposed to obtain an air quality improvement of A. The height of the cost curve measures

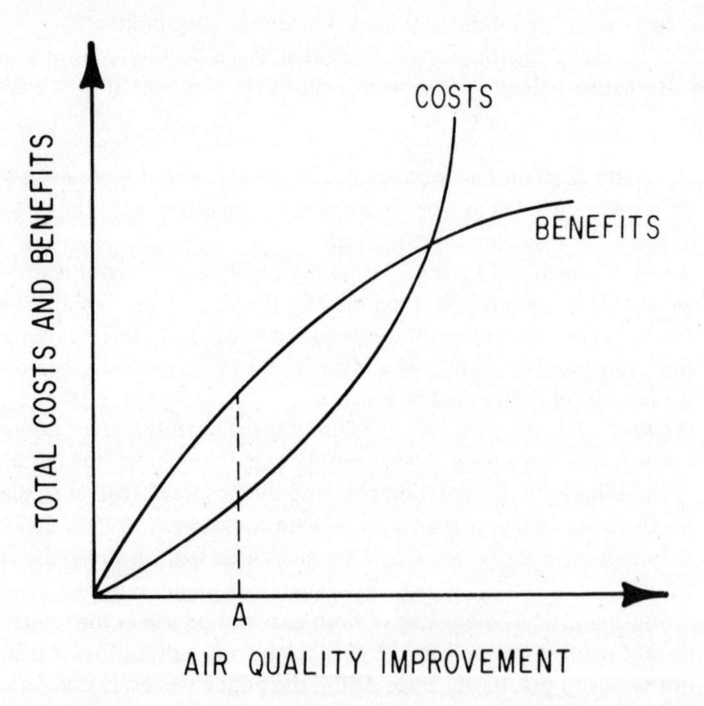

Figure 1-4. Total Benefit and Cost Functions for an Urban Area.

social costs incurred in obtaining the air quality improvement of A. An optimal level of air quality improvement is that level at which the additional social costs and social benefits of a small improvement in air quality (represented by the slopes of the curves in Figure 1–4) are equal. At this point net social benefits–i.e., total benefits minus total costs–are maximized. Within this analytical framework, the problem is to determine an optimal air quality improvement that can be obtained from controls placed on residences. This is equivalent to searching for the most beneficial residential fuel policy.

A benefit-cost analysis is necessarily dependent on predicted values of the variables that affect benefits and costs, a characteristic shared by this study. Additional characteristics of this study that are not as common in other benefit-cost analysis or are unique to this analysis are summarized below.

1. Changes in market shares of space heating fuels are explicitly considered. Some of the market changes are exogenous–i.e., independent–and some are endogenous–i.e., dependent on the policies being analyzed.
2. The analysis reflects private economic decisions based on profit motives (e.g., a landlord chooses to convert to other fuels or to abandon his building).
3. Results are presented for a wide range of fuel policies and market conditions. Therefore, reliance on a single point estimate is not required. In effect, several points on the total cost and benefit curves illustrated in Figure 1–4 are estimated by considering different fuel policies. Furthermore, for each of these policies the study presents a range of assumptions about future prices, housing abandonments, and the like, under which the policy would still be justified.
4. The objective of the study necessitates the assignment of dollar values to variables (e.g., extension of life) that are sometimes considered to have infinite prices. If these variables are evaluated differently by a decision-maker or by society, this analysis can easily be revised.
5. The analysis considers total social benefits and social costs to determine the best fuel policy. It also considers the distribution of the benefits and the costs. In particular, the progressiveness of the policy is analyzed with respect to annual income, racial biases, and effects on property owners, tenants, janitors, fuel distributors, and employees.

It is believed that these five characteristics of this study are essential for making informed policy decisions.

NOTES TO CHAPTER ONE

1. Formerly, this responsibility was delegated to the National Air Pollution Control Administration.

2. United States Code, Vol. 42, p. 1857.
3. Illinois Institute for Environmental Quality and the Illinois Environmental Protection Agency, *State of Illinois Air Pollution Implementation Plan*, Vol. 1 (Argonne, Illinois: Argonne National Laboratory, 1971).
4. U.S. Bureau of the Census, *1970 Census of Housing*, Vol. 1, Part 1, United States Summary (Washington: U.S. Government Printing Office, 1972).

Chapter Two

Social Costs of Residential Fuel Policies

INTRODUCTION

Regulations on the production, sale, purchase, or consumption of any goods are interventions in the functioning of markets. Motivations for regulations are as diverse as reducing monopoly powers, rationing products where shortages exist to provide for equitable distribution, and controlling potential hazards to society. Space heating fuel policies—such as a low sulfur law, a coal ban, or a coal and oil ban—are regulations that affect the energy market. These policies are imposed to obtain social benefits (e.g., improved air quality), which are not considered by the private sector. However, the imposition of the regulations also generates social costs. Regulations are justified only if their social costs are less than their social benefits. The purpose of this chapter is to discuss, identify, and quantify the social costs associated with each of the residential fuel policies discussed in Chapter 1.

The cost to society of a space heating control policy depends on the following:

1. The extent of the usage of each fuel for space heating.
2. The cost of conversion from one fuel to another.
3. The annual costs of operating and maintaining the space heating system.

Because a market consists of suppliers and consumers, any impact on the market affects both. The severity and type of impact is directly related to existing consumption patterns and market trends. If there were no coal users, a coal ban would have no impact on the existing system. On the other hand, if every home was heated with coal, a ban would require substantial capital investments for a distribution system for alternative fuels, in addition to the costs that would have to be borne by the homeowners and landlords. Finally, a coal ban imposed on a declining coal market will have smaller social costs and benefits than one imposed on a stationary or expanding market.

The major expenditures associated with a fuel policy are the costs of conversion to alternative fuels, and the operating and maintenance costs of the space heating equipment. In addition, a landlord has the option to abandon, convert to non-residential use, or demolish his building if the costs associated with an energy policy are greater than the rents forgone because of abandonment, or an alternative use of the property becomes more attractive. Operating and maintenance expenses primarily consist of the cost of the fuel; materials for repairs and upkeep and, in the case of coal, for stoking and cleaning labor. Therefore, labor demands in the form of janitorial services may be affected by a fuel policy.

The above discussion identifies the major parameters affecting the social costs of any space heating fuel policy. In addition, three groups have been identified as being affected by these policies: (1) landlords, homeowners, and renters, (2) fuel distributors and their employees, and (3) janitors. The following sections discuss the nature of social costs in general and some of the difficulties in measuring them; the ways in which energy policies affect fuel distributors, janitorial labor, conversions, and abandonments; the procedures used to estimate the social costs; and the results of the empirical work.

SOCIAL COSTS—A THEORETICAL DISCUSSION[a]

Traditional benefit-cost analyses are related to private industrial development and public investment projects, rather than the regulation of markets. There are no conceptual differences between benefit-cost analyses of market regulations and public investment projects, although in practice, when actual evaluation is needed, they differ. For market policies, costs are borne by large numbers of individuals, at dispersed locations, who are typically non-homogenous in socioeconomic characteristics. For this reason it is important to consider income redistribution effects in addition to aggregate benefits and costs. In addition, the costs of a market policy may be borne over a period of time, sometimes of relatively long duration. These characteristics necessitate the use of more complex and detailed analytical methods than are needed for the evaluation of investment projects. The remainder of this section discusses the issues involved

[a]The theoretical discussion on social costs follows Harberger's three basic postulates: (a) the competitive demand price for a given unit measures the value of that unit to the demander; (b) the competitive supply price for a given unit measures the value of that unit to the supplier; and (c) when evaluating the net benefits or costs of a given action (project, program, or policy), the costs and benefits accruing to each member of the relevant group (e.g., a nation) should normally be added without regard to the individual(s) to whom they accrue.[1] The general characteristics of social cost-benefit analysis are from Mishan.[2] Social costs are measurable only if an economic equilibrium is assumed to exist before and after the imposition of a policy.

in evaluating social costs, with specific reference to the residential fuel policies in Chicago.

For policy evaluation it is imperative to answer the question: What is the *real* cost and benefit *to society* if the policy is imposed? Real costs are imposed on the society if: (1) the policy requires a reallocation of factors of production (e.g., labor, capital) and resources; or (2) factors or resources that would render services without the policy become obsolete and other factors replace them. These are costs since society has to forego resources that otherwise would be used for investment or consumption. It is important not to confuse income redistribution with either costs or benefits unless the redistribution itself generates costs or benefits.

Since the impacts of a fuel policy occur over many years, the related costs or benefits in each year must be combined to allow comparisons between various policies. The procedure used to sum costs over time is a present value calculation. If C_t is the cost occurring in year t, then the present value of these costs over a planning horizon of T years is

$$PV = \sum_{t=0}^{T} C_t/(1+r)^t = A \sum_{t=0}^{T} 1/(1+r)^t, \tag{1}$$

where A is the annualized cost—that is, a constant yearly expenditure which has the same present value as the stream of costs C_t ($t = 0, \ldots, T$)—and r is the social discount rate.

A frequently debated issue concerns the value of the social discount rate.[3] Since the city of Chicago is an open economy, the social discount rate in Chicago is equal to that in the United States. Seagraves argues that this rate is between 8 and 13 percent.[4] A frequently raised argument against a fuel policy is that the discount rate of some landlords is 20 percent or more because they are in a high risk category. Recall that the appropriate discount rate is the social opportunity cost of investment. The high interest rate actually to be paid by the landlords results in an income redistribution and might be a reason for not converting from coal but has nothing to do with the social costs of a fuel policy. In this study a rate of 10 percent is used. Repeating the analysis using a rate of 20 percent does not alter the results.

Another parameter required is the social opportunity cost of labor. This concept can best be explained by an example. If a coal ban is implemented, labor resources in the coal distribution and janitorial sectors would become excessive, while additional labor would be required by the oil and gas companies. If movement from one job to another is perfect—i.e., people displaced from one job obtain another job without any costs—and if fewer people are needed to supply gas and oil heat than were needed to supply coal heat, then the economy would benefit because additional goods and services could be provided by those people freed from the space heating market. However, if labor movement is not perfect (e.g., labor is highly specialized) the reverse might happen and some labor would become unemployed because the "new technology" of space heating would make them obsolete.[5]

For the above example, the latter case is more likely since janitors are highly concentrated in the above 45 age bracket. As a result they would probably not get new jobs but would receive lower wage rates or become underemployed by having to work fewer hours. However, from the landlord's point of view, reduced janitorial service is a saving. Therefore, there is no social cost or benefit associated with less janitorial services, only an income redistribution. This is the approach taken in our analysis. If, in fact, the janitors are not underemployed, then we have overestimated the cost of the policies by ignoring this benefit to society.

Now consider the oil and gas companies that may be required to supply additional goods and services under a space heating control regulation. A representation of the supply curve for oil and gas is given in Figure 2-1. The values Q_0 and Q_1 are the quantities of gas and oil sold in the area without and with a fuel policy, respectively. The area GHQ_0Q_1 is composed of the social cost of the additional gas and oil to Chicago distributors and payments for their services —e.g., storage and labor. From the sum of these two costs should be subtracted

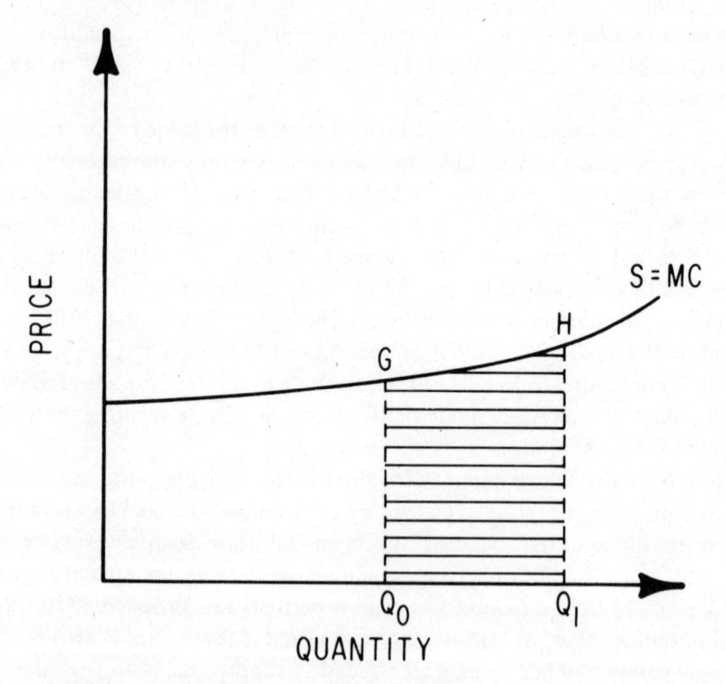

Figure 2-1. Supply Curve of Gas and Oil.

the savings on coal purchases by coal distributors to calculate part of the social cost of the fuel policy associated with fuel conversions: [6]

Social costs = (Change in oil and gas costs to distributors) +
(Change in distribution costs) -
(Change in coal cost to distributors). (2)

Because data were unavailable for these costs the actual computation is different. The sum of the first and second components on the right-hand side of Equation 2 is defined to equal the cost of supplying the additional gas and oil to consumers. Then, instead of subtracting the costs of coal to distributors, we subtracted *the full cost* of coal to residential users. The result is an important figure by itself, it is the difference in fuel costs for space heating due to a conversion from coal to gas or oil. To correct for this excess subtraction, the value of labor services and capital utilized in the distribution of coal are added to the social costs. Making the estimate of social cost in this way provides some insight into the distribution of impacts, which are discussed in detail in Chapter 5.

One major distinction between social and private costs of a policy relates to the prices used. Social costs diverge from private costs when taxes or subsidies are present, a monopoly power is involved, or an externality (e.g., air pollution) in production or consumption is generated. If any of these occur, market prices do not represent social costs. Because the effects of oil and coal consumption on air quality are considered in the benefit analysis in Chapter 4, market prices are not adjusted for this externality.

For gas consumption, the depletion of known gas reserves is another social cost that raises the question of whether the price of gas in Chicago is the social cost of this fuel. Since the price of the gas is regulated, the current (1972) price does not reflect market equilibrium. What is an equilibrium price for gas? Let the equilibrium price be defined as the price that equates the quantity demanded of gas with the quantity supplied while the ratio of reserves to production is maintained at a constant level. A ratio of reserves to production of 20 might be used, because major supply contracts between wells and gas pipeline companies are of 20 years duration. An alternative would be to use a ratio of 13, which was the actual ratio in 1971. [7] This is the definition we used. Hence, the question to answer is by how much should the gas price increase in order to equate supply and demand and maintain the reserve production ratio at 13.

If the demand for gas is price elastic—i.e., sensitive to price changes—any increase in gas prices would lower the quantity demanded; it would also increase exploration activities and thus raise the ratio of reserves to production. If demand for gas is inelastic, then the ratio can only be maintained by increasing reserves. [8] Erickson and Spann estimated a constant price elasticity of new gas discoveries that is not appropriate for this study, since it is only for the short run. [9] Khazzoom estimated long-run relations for new gas discoveries; and revisions and extensions to known reserves.[10] Khazzoom shows that extensions and revisions are more sensitive to ceiling price changes than new gas discoveries. Furthermore, if Alaskan gas discoveries are excluded, the absolute increase in gas reserves due to extensions was about twice as large as the increase from new discoveries in the 1950 to 1970 period.[11] Using the 1971 figures of production, new discoveries, extensions, and revisions of gas, and assuming that (1) new discoveries account for one half of the increase in reserves, and (2) production remains constant at the 1971 level, then the 1969 wellhead ceiling real price of gas would have to increase by about seven cents per thousand cubic feet (mcf) to maintain a reserve production ratio of 13.[12]

The average price paid in 1969 for gas by Peoples Gas, Light, and Coke Company, the natural gas utility servicing Chicago, was 16.6 cents/mcf, and the expenditures on gas purchase were 44.0 percent of total annual expenditures.[13] Therefore, a seven-cent increase in the wellhead price per mcf implies a 42 percent–i.e., 7.0/16.6–increase in the total price paid by Peoples Gas, and an 18.5 percent–i.e., .42 $\times$ .44–increase of the price paid by consumers in Chicago.

The wellhead ceiling price would have to increase continuously to maintain a reserve production ratio of 13, if U.S. gas production continues to increase. If U.S. gas production increases by 500 billion cf. per year (this is about half the annual increase observed in the 1966 to 1971 period),[14] then the price of gas to Chicago consumers would have to increase annually by about 4 percent, in addition to the 18.5 percent immediate increase, to maintain the reserve production ratio at 13.

We have shown that the current price paid by gas consumers in Chicago is below its social price and that if the demand for gas in the United States increases, the social price also increases. In the empirical analysis the January 1972 natural gas price is increased by 50 percent and maintained at this level over the 1973 to 1990 planning horizon. Since an immediate 50 percent increase in the price of gas is greater than an immediate increase of 18.5 percent followed by an annual 4 percent increase, the empirical results somewhat overestimate the social cost of the fuel policies.[15] The January 1972 price of oil was also increased by 50 percent to account for similar externalities associated with oil production and import quotas.

In addition to the social costs related to fuel use, there are three other sources of social costs. The first is the investment in conversions prior to the end of the economic life of the existing capital equipment. The second is the loss of capital that currently provides dwelling services and which, because of abandonment, would have to be replaced by new investments in the residential sector. From the income redistribution point of view, the two costs are levied on the landlords–although some might be transferred on to the tenants. Finally, administration and enforcement costs required for implementing the fuel policies must be considered. These costs would be levied on taxpayers or violators of the regulations who are fined. The following section, which describes the methods used for quantifying the social costs of space heating energy policies, expands on these three cost components.

METHOD FOR QUANTIFYING SOCIAL COSTS

The method described below is a general one and can be applied to any fuel usage policy in any sector. The premise is that when a policy is imposed some individuals must bear costs that would not occur if the policy were not imposed. The model presented below is described with an emphasis on the Chicago residential sector.

The Residential Space Heating Market

As discussed previously, the extent of the usage of each fuel for space heating and how fuel consumption patterns are changing are critical parameters in determining the impacts of a fuel policy. Data for 1962, 1965, 1966 and 1968 to 1972 of the number of dwelling units using coal, oil, gas, and electricity in the city of Chicago were made available by Peoples Gas, Light and Coke Company. These household survey data are disaggregated by type of heating system and building size. The heating system types are (1) group central heating –i.e., a common boiler is used to heat all apartments–(2) individual central heating–i.e., a separate boiler is used to heat each apartment–and (3) room heaters. Building sizes are determined by the number of dwelling units in each building. Twelve building size classes are defined, the smallest being single-family homes and the largest being buildings with 60 or more dwelling units.

Over the ten year period from 1962 to 1972 the total number of dwelling units in the city decreased from 1.18 million to 1.08 million. The market shares of each fuel over that period changed from 35.2 percent to 66.3 percent for gas, 27.5 percent to 14.3 percent for oil, 33.1 percent to 12.0 percent for coal, .1 percent to 1.1 percent for electricity, and 4.1 percent to 6.3 percent for other fuels.

Projections of future fuel market demands were made in three steps.[16] First, the total number of dwelling units in the city was projected by extrapolation of the declining historical trend. It was assumed that the demand for dwelling services would not be affected by space heating control policies. Secondly, projections of the number of dwelling units using each fuel in each building size and for each heating system were made by extrapolating past trends. These latter projections were used to estimate market shares of each fuel for each building size class and type of heating system. Multiplying these market shares by the projected total number of dwelling units provided estimates of the absolute number of living units using each fuel. Finally, if the number of dwelling units using a given fuel in a given building size class was projected to be below the lowest value of the range of dwelling units defining the building size class, then the number of dwelling units was set equal to 0.

This approach was based on the premise that extrapolation of trends for an aggregate is more accurate than the extrapolation of trends of its parts. Furthermore, there were 8 data points available to project total dwelling units and a maximum of only 5 to generate market shares for the low sulfur fuel policy.

Because the city of Chicago enacted a low sulfur law in 1968, which altered the residential space heating market, two trends of the number of dwelling units using each fuel were observed, one for the pre-1968 and one for the post-1968 period. The former reflects a no control situation with relatively stable fuel prices. The latter reflects the low sulfur law and a large increase in the price of coal relative to the increases in the prices of gas and oil.

The pre-1968 trend extrapolated into the future represents a hypothetical situation used as the basis to compare the costs of the sulfur law policy. The economic variable primarily responsible for the difference in pre- and post-1968 trends is the price of coal. The low sulfur law forced coal users to shift from southern Illinois coal to eastern Kentucky coal during the 1968 to 1971 period. Over this period the price of a ton of coal increased by about 100 percent. However, the price of coal nationally also increased drastically during this period because of strip mining and coal mine safety regulations. Therefore, only a portion of the observed coal price increase should be attributed to the low sulfur law. Accordingly, the pre-1968 trend was adjusted to account for the price increases that are independent of the policy.

Let Figure 2-2 represent the pre-1968 extrapolated trend and the post-1968 data of coal-using dwelling units. Assume that the price of coal was 20

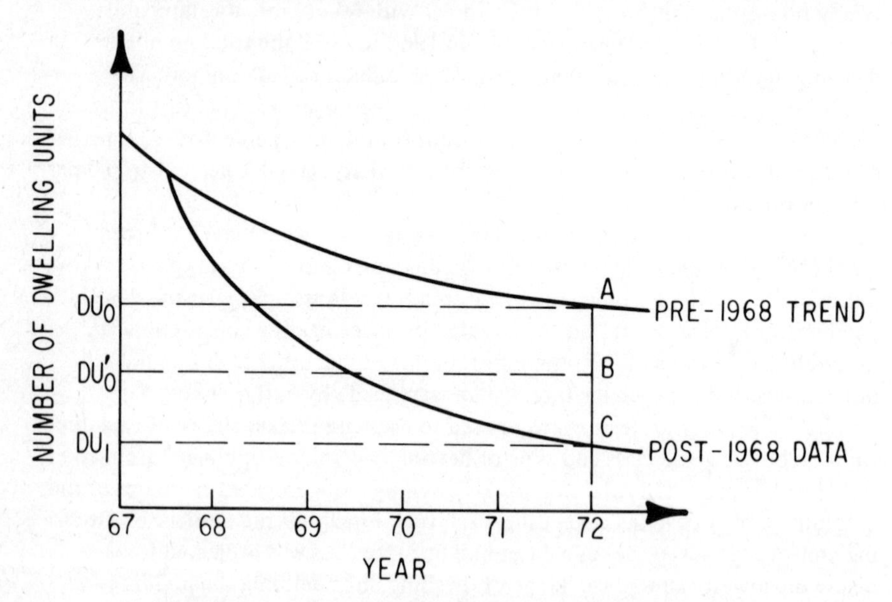

Figure 2-2. Number of Coal-Using Dwelling Units.

dollars per ton in 1968, 33 dollars per ton in 1972, and that the price of gas and oil remained constant at their 1968 levels. Then in 1972 there would have been DU_0 dwelling units if the price of coal remained around 20 dollars per ton (assuming no inflation), and there would be DU_1 dwelling units if the price changed to around 33 dollars per ton. If the price of coal would have changed to 30 dollars per ton without a sulfur law because of other regulations recently imposed on the coal industry, then the pre-1968 or no control case would have resulted in DU_0' coal-using dwelling units in 1972. In the calculations we assume that the pre-1968 trend shifts in direct proportion to the ratio of the estimated no control policy price difference of coal to the observed price difference–i.e., $AB/AC = (30\text{–}20) / (33\text{–}20) = .77$.

The coal ban and the coal and oil ban policies were simulated using the post-1968 projections and forcing the complete conversion from coal or from coal and oil by 1975 or 1977. A linear conversion path was assumed. Projected market shares of oil, natural gas, and electricity (only natural gas and electricity for the coal and oil ban) were used to estimate conversions to those fuels by coal or by coal and oil users. The conversion shares used are the 1980 extrapolated fuel market shares using the post-1968 data. These projections assume that present fuel prices will prevail and that there will still be a small shortage of natural gas in Chicago. (A waiting list for gas attachments was recently initiated in Chicago.) These market shares are presented in Table 2-1. Note that there are different market shares for residential buildings with group central heating plants, individual central heating plants, and room heaters. Conversions to electricity were found to be economical, even under the most favorable conditions, only for people using room heaters. Since only 700 dwelling units used coal room heaters and only 1500 used oil room heaters in 1972, conversions to electricity were not considered in this study.

Fuel Conversions and Abandonments

Using the procedure described above, the annual change in the number of dwelling units using a given fuel can be estimated.[17] This change is the net effect of conversions from one fuel to another, abandonment or demolition

Table 2-1. Fuel Conversion Shares (Percent)

	GCH	*ICH*	*RMH*
Coal Ban			
Oil	11	8	0
Gas	89	92	87
Electricity	0	0	13
Oil Ban			
Gas	100	100	87
Electricity	0	0	13

of buildings, and the construction of new buildings. Since the fuel policies affect coal and oil users and new buildings in Chicago are almost exclusively heated by natural gas or electricity, it is reasonable to assume that the observed decreases in dwelling units using coal and oil are the net result of conversions and abandonments. It was also assumed that the rate of conversion, plus abandonment from oil, was not affected by the sulfur law. Therefore, the observed decrease in the rate of decline in oil-using units after 1968 was assumed to result from an increase in the number of coal users converting to oil. Finally, a "natural" abandonment rate under the low sulfur law was estimated by assuming that the observed decrease in the total number of dwelling units in Chicago over the 1968 to 1972 period resulted from abandonments and that half of these were coal-using and half were oil-using dwelling units. A 5 percent abandonment rate for both coal- and oil-using dwelling units was found to be a good approximation for the true abandonment rate. The number of conversions is equal to the difference between the total change and the number of abandonments of coal- or oil-using dwelling units.

Estimates of the costs of conversions were made by multiplying the number of conversions by the unit cost. Conversion costs per dwelling unit were estimated by building size and heating system to account for economics of scale. Conversion costs were defined as the cost of altering the furnace to burn another fuel and the costs of an on-site storage tank or pipe line needed to supply the new fuel. Some homeowners or landlords may decide to install a new furnace rather than modify their old system. It can be assumed that they would do so only if the returns are greater than the returns from a modification, since the policy does not require the installation of a new furnace. Therefore, the social cost of the policy should not include this extra expense. In other instances, costs for improving the plumbing, electricity, or other unrelated repair items might be required to obtain a permit to make the fuel conversion. Inasmuch as these repairs result in improved (safer) housing services to tenants, they do not contribute to net social costs. Thus, these costs were not considered to be part of the social costs of a fuel policy. On the other hand, one might regard these costs as part of the social costs, but then the improved dwelling services would also have to be considered as part of the benefits. Therefore, there is no net contribution to social costs.

The conversion cost estimates per dwelling unit for each building size class were obtained by using representative building characteristics and engineering assumptions. Wage rates for the Chicago area were used to estimate the contractors' labor charges. Furthermore, it was assumed that the supply of conversions (contractors) is infinitely elastic—i.e., conversion costs will not change with demand. Therefore, conversion costs were not affected by the control policies. Finally, actual conversion costs for a few buildings in Chicago were available. These data were used to adjust the synthetic cost estimates so they would better reflect actual conversion costs. The unit conversion costs used in the analysis are provided in Table 2-2.

Table 2-2. Conversion Costs per Dwelling Unit (Dollars)

Building Size Class	Average Number of Dwelling Units	Conversions to Oil			Conversions to Gas		
		GCH[a]	ICH[b]	RMH[b]	GCH	ICH[b]	RMH[b]
1	1	600	600	600	510	510	510
2	2	500	600	600	425	510	510
3	3	450	500	500	383	425	425
4	4	400	500	500	333	425	425
5	6	350	500	500	300	425	425
6	10	300	500	500	255	425	425
7	16	238	500	500	209	425	425
8	22	195	500	500	170	425	425
9	26	177			154		
10	34	150			129		
11	47	126			111		
12	133	100			92		

[a]GCH ≡ Group central heating; ICH ≡ individual central heating; RMH ≡ Room heaters.
[b]There are no buildings with this type of heating plant in classes 9 through 12.

A building may be abandoned or demolished rather than having the boiler converted if it is "marginal." A marginal building is usually in poor condition and requires major upgrading when converted to a new fuel while the rental income is fixed or declining. Also, because of neighborhood conditions, the expected life span of most of these buildings as a source of residential services is relatively short. If a building is abandoned, then the cost to the landlord is the value of the rents lost, which is assumed to be 300 dollars per dwelling unit per year. (This assumes that the landlord has no alternative use of the residential property.) The tenants displaced may also bear costs. These are the costs of searching for new lodging, moving, and inconvenience. The costs to tenants are accounted for in the analysis by assuming search and moving costs to be 100 dollars, and the value of inconveniences to be 100 dollars for three years after they were forced to move.[18] Note that any increased rent payment is not a social cost but rather an income redistribution. Both the costs to the landlords and tenants are social costs because they require the diversion of factors of production and resources from other uses to the residential sector.

Two parametric analyses related to conversions and abandonments were conducted. The first assumes that all the coal (coal and oil) dwelling units would convert to cleaner fuels under a coal (coal and oil) ban. The second assumes that all of the dwelling units that would have been abandoned in the 1975 (1977) to 1990 period–assuming a 5 percent abandonment rate–would be prematurely abandoned in 1975 (1977). (The second analysis was only conducted for the coal ban policies.)

If the sulfur law continues, 84 thousand coal-using dwelling units will be left in 1975 (61 thousand in 1977). Assuming a 5 percent annual abandonment rate, 25.7 thousand of these dwelling units will be abandoned by 1990 (17.9 between 1977 and 1990). Therefore, when abandonments are considered in the analysis, it is assumed that a massive abandonment of 25.7 thousand dwelling units will occur in 1975 (17.9 in 1977). Finally, it is assumed that the displaced tenants find new lodging within their present neighborhood and that the resulting fuel consumption pattern in 1975 (1977) is the same when abandonments or conversions take place.

When the abandonment assumptions are employed, the cost calculations are affected in a number of ways. First of all, the costs of abandonment must be substituted for the costs of conversion for those dwelling units that are abandoned. Secondly, the estimated number of coal-using dwelling units in the 1972 to 1975 (1972 to 1977) period is altered (see Figure 2-3). This affects all of the cost and benefit calculations for that period.

When abandonments are not considered, both the sulfur law and coal ban curves are assumed to represent conversion only. When abandonment is considered, the sulfur law curve is assumed to indicate the combined effect of a 5 percent abandonment rate and conversions. Because the assumed linear path under the coal ban also includes these abandonments, conversions do not follow a linear path. Rather, the numbers of conversions that would take place

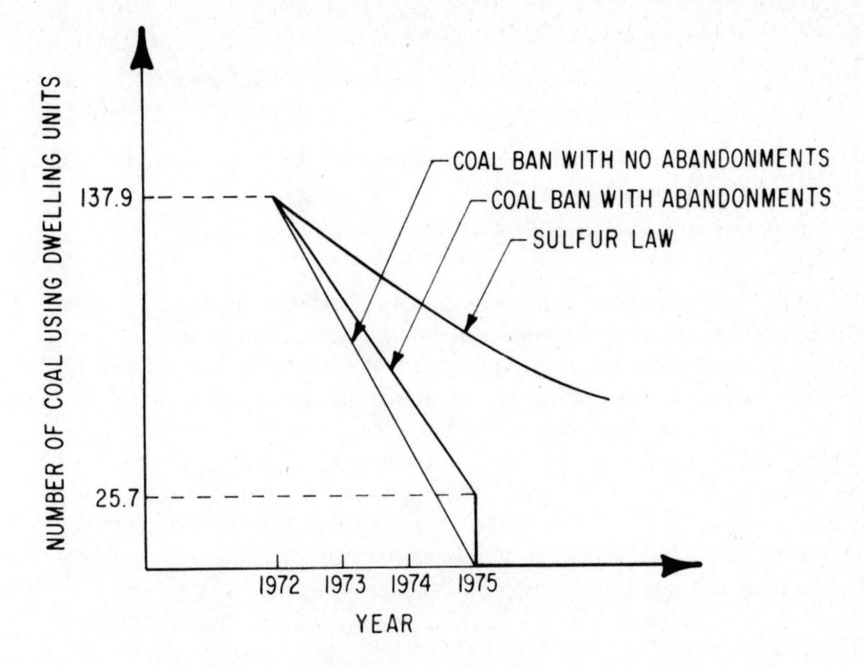

Figure 2–3. Conversion Paths Resulting from a 1975 Coal Ban.

under these assumptions are 30.5, 31.6, and 32.4 thousand for 1973, 1974, and 1975, respectively (17.1, 18.2, 19.0, 19.8, and 20.4 for 1973 through 1977).

Operating and Maintenance Costs

The major expense of providing heat is the fuel cost. The labor and capital costs for operating and maintaining oil, gas, and electric space heating units are minimal and approximately the same. The labor costs for coal units are not considered for reasons stated in the discussion on the opportunity cost of labor. Therefore, we shall only consider fuel costs in this section.

The difference between the costs of heating with various fuels resulting from fuel conversions is either a savings or expenditure to homeowners or landlords. In this study these differences are employed as a measure of the social costs of diverting resources from the importing and distributing of dirty fuels to clean fuels.

Given the number of dwelling units using each fuel, the energy costs in year t are calculated as follows:

$$F_{it} = \sum_{j,k} (DU_{ijkt})\,(Q_{ijk})\,(P_{it}), \qquad (3)$$

where

F_{it} ≡ the annual cost of fuel i in year t;

DU_{ijkt} ≡ the number of dwelling units using fuel i in buildings of size j with heating system k in year t;

Q_{ijk} ≡ the quantity of fuel i consumed per dwelling unit in buildings of size j with heating system k;

P_{it} ≡ the unit price of fuel i in year t.

The quantity of a fuel consumed per dwelling unit is related to the efficiencies of the boilers, the number of exterior walls, the degree of temperature control, and the heat value (Btu rating) of the fuel. Boilers commonly used for residential space heating become more efficient with size. Therefore, the per dwelling unit consumption of a fuel will decrease with the size of a building that has group central heating. The per dwelling unit consumption of fuels also decreases with the size of a building (with any type of heating plant) because fewer walls are exposed to the outside. In very large buildings, most apartments have only one exterior wall (corner apartments have two) and only the top floor will lose heat through the ceiling. On the other hand, most single-family homes have at least four exterior walls and also lose heat through the ceiling.

Many older buildings with group central heating do not provide individual temperature controls in each living unit. Therefore, many tenants who like a cooler temperature will open windows to adjust the heat. This situation becomes more likely as the building size increases because of the increase in the number of tenants and the likelihood that many apartments will become excessively hot when the boiler control is set to heat properly the coldest dwelling unit. This results in increased per dwelling unit consumption with size.

The combination of these effects is simulated in the energy cost model described in Appendix C. The efficiencies and heat loss effects are based on engineering approximations. The effects of temperature control are modeled by altering the load factor of the boiler. The per dwelling unit fuel consumption rate changes in different directions for each of these parameter changes as the size of the building increases, thus the aggregate change is not always in one direction. This is shown in Table 2-3, which provides the basic fuel consumption parameters used in the study.

All of the above parameters affecting energy consumption are characteristics of the housing stock and are independent of fuel policies. However, the characteristics of the fuel used also affect the quantity consumed. If X Btu's are required to heat an apartment each year and a fuel has Y Btu's per unit, then X/Y units of the fuel will be consumed. The amount of fuel required is reduced as the value of Y is increased.

The heat value of coal was affected by the Chicago low sulfur law. Prior to the passage of the low sulfur law, southern Illinois coal with an average

Table 2-3. Annual Fuel Consumption per Dwelling Unit[a]

Building Size Class	Average Number of Dwelling Units	Tons of Coal			Gallons of Oil			Cubic Feet of Gas		
		GCH[b]	*ICH*[c]	*RMH*[c]	*GCH*	*ICH*[c]	*RMH*[c]	*GCH*	*ICH*[c]	*RMH*[c]
1	1	11.42	11.42	11.42	1260	1260	1260	156,000	156,000	156,000
2	2	10.40	11.42	11.42	1150	1260	1260	143,000	156,000	156,000
3	3	8.93	10.40	10.40	990	1150	1150	123,000	143,000	143,000
4	4	7.89	10.40	10.40	880	1150	1150	109,000	143,000	143,000
5	6	6.69	10.40	10.40	752	1150	1150	93,000	143,000	143,000
6	10	5.72	10.40	10.40	655	1150	1150	81,000	143,000	143,000
7	16	5.35	10.40	10.40	625	1150	1150	78,000	143,000	143,000
8	22	5.34	10.40	10.40	638	1150	1150	80,000	143,000	143,000
9	26	5.41			658			82,000		
10	34	5.29			625			82,000		
11	47	4.74			620			77,000		
12	133	3.56			509			68,000		

[a]Heat value of coal is 23×10^6 Btu/ton; of oil $13.9 \times 10^4 - 15 \times 10^4$ Btu/gal; of gas 10^3 Btu/cu. ft.
[b]GCH ≡ Group central heating; ICH ≡ Individual central heating; RMH ≡ Room heaters.
[c]There are no buildings with this type of heating system in classes 9 through 12.

heat value of 23×10^6 Btu/ton was consumed in Chicago. The law caused a switch to eastern Kentucky coal. This coal, in addition to having a lower sulfur content, also has a higher average heat value of 26×10^6 Btu/ton. Therefore, less coal is required per dwelling unit per year under the low sulfur law than under the no control situation. The opposite situation is possible for oil. Distillate oil can be mixed with residual oil to reduce the sulfur content. This would result in an increase in the consumption of oil because distillate oil has a lower average heat value than residual oil. Since most residential buildings cannot use No. 6 residual oil and only very large buildings can use No. 5 residual oil, this mixing was not widespread and was not considered in our analysis.

The effect of the sulfur law on heat values is the only policy impact on the per dwelling unit consumption of fuels that was considered in the study. The fuel ban policies affect Equation 3 by altering the DU and P parameters. The effects on the number of dwelling units were discussed above. The impact of the space heating control policies on prices is the topic of the following discussion.

Two types of price changes are considered: (1) exogenous changes–i.e., those that occur independently of the fuel policies–and (2) endogenous changes–i.e., those directly resulting from a control policy. The distinction between the two is quite important when the benefits and costs of the policies are compared. For example, an endogenous price increase in natural gas resulting from a coal and oil ban affects the costs of the policy by changing the estimated fuel costs. Because coal and oil users are forced by the regulation to convert to gas or electricity, the price of gas does not affect the number of conversions. In addition, this price increase does not affect the rate of conversions assumed for the sulfur law since it is assumed to result from the ban. Therefore, benefits are not affected. On the other hand, exogenous price changes will affect the conversion rates of coal and oil users under the sulfur law and alter the relative (to the sulfur law) improvements in air quality resulting from a fuel ban. Thus, both the costs and benefits change.

Endogenous and exogenous price changes differ in another important way. Since endogenous price changes, by definition, are caused by a control policy, the changes in fuel costs for everyone in the city must be attributed to the control policy. Although exogenous price changes also change fuel costs for everyone, these changes cannot be attributed to a fuel policy since they would occur with or without the policy. Exogenous price changes only affect the costs of a fuel policy by altering the fuel saving or costs resulting from fuel conversions.

Exogenous price change effects are modeled in the same way that the no control trend was adjusted to account for coal price increases due to strip mining and mine safety regulations. In fact, these are by definition exogenous price changes. Now the interest is in increases in oil and gas prices relative to coal prices in the post-1968 period. Such increases would result in an upward shift in the sulfur law trend curve (see Figure 2-2) because the rate of conversion away from coal would decrease.

An endogenous change in the price of a clean fuel, say natural gas, resulting from a fuel policy would occur if the supply curve looked like Figure 2-1—i.e., less than perfectly elastic. Space heating demands occur in the winter requiring the gas company to store gas during the summer to meet winter peaks. The underground storage facility which serves Chicago is being used to capacity. Therefore, liquid storage facilities or marginal underground facilities may be required to handle the increased demand for gas resulting from a fuel policy. The recent addition of a liquid storage facility for Chicago and the inability of the gas company to satisfy existing demands seem to support the hypothesis that endogenous gas price increases will occur because of additional storage costs.

A winter supply curve of gas is given in Figure 2-4 to facilitate the discussion of the method used to estimate the endogenous price increases. The demand for gas will shift upward as the number of conversions to gas increases. (Recall that the gas utility is regulated, and thus the rate of profit it is allowed to make is limited. For simplicity assume that the supply function the consumers are facing is already adjusted for this constraint on returns to capital.) As demand shifts from D_0 to D_1 the price increases from P_0 to P_1. The relative increase in the quantity demanded is

$$(Q_1 - Q_0)/Q_0 = \Delta Q/Q_0 \tag{4}$$

and the relative increase in the price is

$$(P_1 - P_0)/P_0 = \Delta P/P_0. \tag{5}$$

The elasticity of supply is defined as

$$\eta = (\Delta Q/Q_0/(\Delta P/P_0). \tag{6}$$

We assume the supply elasticity is constant over the range $Q_0 - Q_1$. Therefore, if η and the relative change in the quantity demanded are known, the relative increase in price can be calculated as follows:

$$\Delta P/P_0 = (\Delta Q/Q_0)/\eta. \tag{7}$$

For any fuel policy the increase in the demand for gas is estimated. We assume supply elasticities of 3 and 5 to estimate possible gas price increases resulting from these policies. Because these elasticities are greater than one, the supply curve in Figure 2-4 does not depict these elasticities; rather, the curve would have to increase at a decreasing rate. Finally, given these price

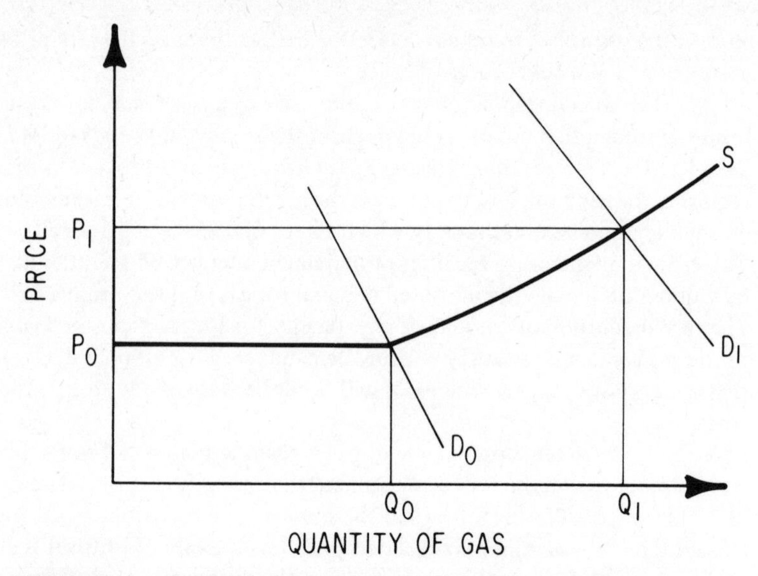

Figure 2-4. Gas Price Changes Due to a Demand Shift.

increases, they must be applied to all gas consumers, not just those converting to gas.

Endogenous price changes for oil were assumed not to occur. Therefore, the price of gas relative to oil would increase, making conversions to oil more attractive. However, conversions to oil under a coal ban were assumed to occur because of the natural gas shortages in Chicago. If there were no shortages, then all coal users who would convert rather than abandon would switch to gas because of the lower conversion cost (no on-site storage tank is required) and the presently lower relative gas price. Gas prices would have to increase by about 40 percent and oil prices remain constant, before conversions from coal to oil would be more attractive than conversion to gas. Therefore, the market shares in Table 2-1 for conversion to oil and gas under a coal ban were assumed to be unaffected by the endogenous price increases in natural gas. The basic fuel cost data are presented in Tables 2-4 and 2-5.

Endogenous price changes as described above occur because the local demand for gas increases, and the local gas distributors' long-run winter supply curve was assumed to be upward sloping. Since the demand for gas under the low sulfur law would eventually be the same as the demand under any of the fuel ban policies, the costs associated with endogenous price increases occur prematurely with a fuel ban policy. Therefore, the procedure for calculating these costs is similar to the procedure used to estimate conversion and abandonment costs. It is interesting to note that from 1972 to 1990 the estimated price increase resulting from a sulfur law would be the same as the increase due to the 1975 coal ban between 1972 and 1988. (While these data assume an elasticity

Table 2-4. Annual Coal Costs per Dwelling Unit ($/Yr/DU)

Building Size Class	Average Number of Dwelling Units	Season[a] Btu/Ton $/Ton	1969–70 23×10^6 16.75[b]	1970–71 23×10^6 17.50[c]	1970–71 23×10^6 22.60[d]	1970–71 23×10^6 25.00[b]	1971–72 23×10^6 18.28[c]	1971–72 23×10^6 27.05[d]	1971–72 26×10^6 33.35[b]
1	1		191	200	258	285	208	309	337
2	2		174	182	235	260	190	281	307
3	3		150	157	202	223	164	242	263
4	4		132	140	178	197	144	213	233
5	6		112	117	151	167	122	181	197
6	10		96	100	130	143	105	155	169
7	16		90	94	121	134	98	145	158
8	22		89	93	120	133	97	144	157
9	26		91	95	123	136	99	147	160
10	34		89	93	120	133	97	144	157
11	47		79	83	107	118	86	128	140
12	133		60	63	81	90	65	97	105

[a] For the heating seasons 1970–71 and 1971–72 three values are given. The first is for the observed pre-1968 trend—i.e., no control case—the second for the adjusted no control case, and the third for the sulfur law—i.e., the post-1968 trend.

[b] Observed market prices.

[c] Price of Illinois coal is assumed to increase by the same rate as coal prices increased nationally in the 1964 to 1968 period.

[d] Price of Illinois coal is assumed to increase by the same rate as coal prices increased nationally in the 1969 to 1971 period.

Table 2-5. Annual Gas and Oil Costs per Dwelling Unit ($/Yr/DU)

Building Size Class	*Average Number of Dwelling Units*	*Natural Gas*			*Oil*		
		1969-70	*1970-71*	*1971-72*	*1969-70*	*1970-71*	*1971-72*
1	1	178	185	192	210	222	228
2	2	150	156	163	192	203	208
3	3	125	130	136	166	174	179
4	4	109	114	118	147	155	159
5	6	91	95	99	122	129	133
6	10	76	80	84	106	112	115
7	16	71	74	78	101	107	110
8	22	71	75	78	104	109	112
9	26	73	77	80	107	113	116
10	34	72	75	79	107	113	116
11	47	66	69	73	100	106	109
12	133	55	58	61	75	84	87

Note: The prices used are for either December or January for each heating season.

of 3, a similar result is obtained with an elasticity of 5.) However, with a coal and oil ban this price level would be exceeded by 1975. The impact of this is demonstrated by the results that follow.

RESULTS OF COST CALCULATIONS

Fuel Policy Costs with Constant Fuel Prices

The following results are obtained from the summary cost tables in Appendix F. The costs presented are the additional costs of the policy (marginal costs) relative to a background situation. In the summary tables of Appendix F, additional results obtained by altering our assumptions–i.e., parametric analyses–are presented.

1. *The low sulfur law.* To calculate the marginal costs of the low sulfur law, it was compared to the policy of no control that was simulated by adjusting the 1962 to 1968 trend. The adjustment reduced the apparent impact of the sulfur law on the rate of conversions from coal by 78 percent (see Appendix F). The marginal annualized cost to be attributed to the low sulfur law is $.09 million for a 10 percent discount rate and minus $.11 million for a 20 percent discount rate–i.e., a savings.

2. *A coal ban by 1975.* Under most assumptions this policy was cost free (relative to the sulfur law) when fuel prices were held constant. The reason is the large saving on fuel costs that dominates all other cost components. For a 10 percent discount rate the calculated annualized saving is $1.39 million. This saving is reduced to $.69 million when a 20 percent discount rate is used.

3. *A coal ban by 1977.* The results are similar to the 1975 coal ban, except that the net savings are smaller. They are, respectively, $1.01 million and $.62 million with the two discount rates.

4. *A coal and oil ban by 1975.* The cost figures presented in Appendix F for the coal and oil ban are relative to the low sulfur law. The marginal cost of banning oil in addition to coal is calculated by subtracting the cost of the coal ban from this total. Using a discount rate of 10 percent still renders a net savings for the coal and oil ban, although smaller than the savings of a coal ban by itself ($.97 vs. $1.39 million per year). Hence, the net cost of banning oil in addition to coal is $.42 million annually. A discount rate of 20 percent increases this cost to $2.34 million.

5. *A coal and oil ban by 1977.* Changing the date of compliance for the coal and oil ban reduces its net annualized costs to $.28 and $2.09 million, respectively.

The fuel policy cost estimates presented above assume that no buildings will be abandoned as a result of the policy. The effects of abandonment on policy costs were considered for the 1975 and 1977 coal bans. The rents lost by landlords–i.e., the value of the lost residential resources–and the moving and psychic costs borne by the displaced tenants are presented in Table 2-6.

Table 2-6. Components of Abandonment Costs (Millions of Dollars—1973 Present Values)

	Coal Ban	*Sulfur Law*	*Differences*	*Coal Ban*	*Sulfur Law*	*Differences*
Discount Rate		*10%*			*20%*	
1975 Coal Ban (1975 to 1990 Period)						
Rent Lost	54.83	34.63	24.20	30.39	16.22	14.17
Moving Expenses	2.12	1.50	.62	1.78	1.00	.78
Psychic Costs	5.81	4.08	1.73	4.51	2.52	1.99
Total Cost	62.76	40.21	26.55	36.68	19.74	16.94
Annualized Total Cost	6.97	4.46	2.95	6.35	3.42	2.93
1977 Coal Ban (1977 to 1990 Period)						
Rent Lost	29.72	18.88	10.84	14.33	7.75	6.58
Moving Expenses	1.22	.89	.38	.86	.50	.36
Psychic Costs	3.34	2.39	.95	2.18	1.26	.92
Total Cost	34.28	22.16	12.12	17.37	9.51	7.86
Annualized Total Cost	3.81	2.46	1.35	3.01	1.65	1.36

The results are marginal with respect to the sulfur law under which a 5 percent abandonment rate was assumed.

Because the assumed paths of conversions are different for a coal ban with and without abandonments, the abandonment costs presented in Table 2-6 cannot be added to the previous cost estimates to obtain the total cost of the coal ban relative to the sulfur law. Including abandonments in the analysis results in the reduction of the losses to coal distributors, conversion costs, and fuel savings. However, the simple addition of the abandonment cost results in a good approximation of the detailed cost calculation because these additional effects tend to cancel each other. For example, the detailed cost analysis for the 1975 coal ban resulted in an annualized cost of $1.50 million using a 10 percent discount rate. The simple addition of the $1.39 million savings assuming no conversions and the $2.95 million abandonment costs provides an estimate of $1.56 million, a slight overestimate. Similar results occur for the 1975 coal ban with a 20 percent discount rate and the 1977 coal ban with both discount rates. The detailed cost estimates are $2.03, $.41, and $.71 million, respectively. (See Tables F-9 and F-10 for the breakdown of these costs.)

Finally, the administration and enforcement costs associated with each policy must be added to these costs because they represent the value of the resources (human and non-human) that society must reallocate to ensure that the regulations are implemented. The city of Chicago's Department of Environmental Control plans to spend about $.5 million on enforcement in fiscal year 1974.[19] This cost includes all enforcement activity for air, water, noise, etc., in Chicago. If this is used as an estimate of both the enforcement and administrative costs of the low sulfur law, then it is likely that these costs are overestimated. It will also be assumed that no additional costs will be needed to implement the other residential fuel policies—i.e., the marginal administration and enforcement costs are 0. This is based on the hypothesis that short-term legal costs associated with a fuel ban (e.g., the coal merchants' suit to invalidate the state's de facto coal ban) will be offset by lower enforcement costs compared with the sulfur law. (It is easier to determine whether someone is or is not using coal than to determine whether or not he is using low sulfur coal.) Based on these assumptions the annualized cost of the low sulfur law is increased by $.5 million. Table 2-7 summarizes the marginal costs of the five policies with constant fuel prices.

Fuel Policy Costs with Changing Fuel Prices

In Table 2-8 the marginal effects of an exogenous increase in oil and gas prices of 50 percent in 1973 are summarized.[20] Two coal-use adjustment factors for the sulfur law are considered; they are .768 and .406. The adjustment factor is used to reduce the number of conversions from coal with a low sulfur law because the increase in the price of gas and oil makes conversion less attrac-

Table 2-7. Marginal Costs of Fuel Policies Assuming Constant Fuel Prices (Millions of Dollars)

Discount Rate	*10%*	*20%*
Low Sulfur Law vs. No Control	.59	.39
Coal Ban 1975 vs. Low Sulfur Law	(1.39)	(.69)
Coal Ban 1977 vs. Low Sulfur Law	(1.01)	(.62)
Coal & Oil Ban 1975 vs. Coal Ban 1975	.42	2.34
Coal and Oil Ban 1977 vs. Coal Ban 1977	.28	2.09

Note: The values shown are annualized costs. Parentheses indicate negative costs or savings. Fuel prices are constant at the 1971 level. (See Appendix F for other parametric changes.)

Table 2-8. Additional Costs to the Residential Sector Due to a 50 Percent Exogenous Increase in Gas and Oil Prices (Millions of Dollars)

Discount Rate	*10%*	*10%*	*20%*	*20%*
Adjustment Factor for Coal	*.406*	*.768*	*.406*	*.768*
Coal Ban 1975 vs. Low Sulfur Law	1.33	1.31	1.92	1.11
Coal Ban 1977 vs. Low Sulfur Law	1.00	.95	1.55	1.27
Coal and Oil Ban 1975[a] vs. Coal Ban 1975	(2.04)	–	(2.24)	–
Coal and Oil Ban 1977[a] vs. Coal Ban 1977	(1.65)	–	(1.60)	–

[a]No change is assumed in the rate of conversions away from oil–i.e., the adjustment factor for oil conversion is 1. The values shown are annualized costs. Parentheses indicate negative costs or savings.

Source: Appendix G.

tive. The results for the coal and oil ban are only tentative but are within the order of magnitude of the real costs.

In Table 2-9 the marginal effects of endogenous price increases of natural gas caused by the coal bans and the coal and oil bans are summarized. Two alternative elasticities of winter gas supply are assumed; they are 3 and 5. Recall that with an endogenous price change, the fuel cost increases to all gas consumers, including existing users, are social costs of the policy. Since the residential sector uses approximately half of the winter gas consumed in

Table 2-9. Additional Costs to the Residential Sector Due to Endogenous Increases in Gas Prices (Millions of Dollars)

Discount Rate	*10%*	*10%*	*20%*	*20%*
Elasticity of Gas Supply	*3*	*5*	*3*	*5*
Coal Ban 1975 vs. Low Sulfur Law	.65	.42	.72	.45
Coal Ban 1977 vs. Low Sulfur Law	.45	.30	.45	.29
Coal and Oil Ban 1975 vs. Coal Ban 1975	3.34	2.02	3.37	2.07
Coal and Oil Ban 1977 vs. Coal Ban 1977	2.76	1.65	2.61	1.57

Note: The values shown are annualized costs.

Chicago, the cost to the entire Chicago economy of an endogenous price increase of gas is twice the cost to residences.

Because exogenous price changes affect the rates of decline in coal- and oil-using dwelling units under the low sulfur law, and endogenous price increases are based on absolute price levels–i.e., a 50 percent increase in the base price of gas would increase the endogenous price increases by 50 percent–the effects of simultaneous endogenous and exogenous price changes would not be equal to the sum of the individual effects. A better approximation of these costs is achieved by multiplying the costs in Table 2-9 by 1.5 and adding these to the corresponding costs in Table 2-8.

Since Tables 2-8 and 2-9 provide cost estimates that are additional or marginal costs relative to those provided in Table 2-7, estimates of policy costs under various price change assumptions can be made. For example, consider the following set of assumptions: (1) an exogenous gas and oil price increase of 50 percent will occur with an adjustment factor of .768; (2) the social discount rate of 10 percent is applicable; and (3) the conversion costs are those provided in Table 2-2. The marginal cost of the 1975 coal ban under these assumptions would be (-1.39 + 1.31) = -.08 or a savings of $.08 million a year. Cost calculations using additional parametric changes can be made using these tables and those presented in Appendix F.

Table 2-10 shows the cost estimates for each policy assuming: (1) no exogenous price increase will occur; (2) an endogenous gas price increase will occur and the appropriate elasticity of winter gas supply is 3; and (3) the conversion costs are those provided in Table 2-2. These are the assumptions believed to be the most plausible among all the alternatives considered. They will be referred to as the basic assumptions. The cost estimates in Table 2-10 will be used for the initial comparison with benefits in Chapter 6.

Table 2-10. Summary: Costs of Fuel Policies Under the Most Likely Assumptions (Millions of Dollars)

Policy Comparison	*Estimation Period*	*Discount Rate*	
		10%	*20%*
Sulfur Law vs. No Control	1969–1990	.59	.39
Coal Ban 1975 vs. Sulfur Law	1973–1990	(.74)	(.27)
Coal & Oil Ban 1975 vs. Coal Ban 1975	1973–1990	3.76	5.71
Coal Ban 1977 vs. Sulfur Law	1973–1990	(.56)	(.17)
Coal & Oil Ban 1977 vs. Coal Ban 1977	1973–1990	3.04	4.70

Note: Parentheses indicate negative costs or savings.
Source: Tables 2–7 and 2–9.

NOTES TO CHAPTER TWO

1. A.C. Harberger,"Three Basic Postulates for Applied Welfare Economics, An Interpretive Essay," *Journal of Economic Literature*, Vol. 9, No. 3 (1971), pp. 785–797.
2. E.J. Mishan, *Cost Benefit Analysis* (London: George Allen and Unwin, Ltd., 1971).
3. See, for example, A.C. Harberger, "On Discount Rates for Cost Benefit Analysis," in *Project Evaluation*, ed., A.C. Harberger (Chicago: Markham Publishing Co., 1973), pp. 70–93; W.J. Baumol, "On the Social Rate of Discount," *American Economic Review*, Vol. 58, No. 4 (1968), pp. 788–802; A. Sandmo and J. Dreze, "Discount Rates for Public Investment in Closed and Open Economies," *Economica*, New Series, Vol. 38, No. 152 (1971), pp. 395–412; and J.A. Seagraves, "More on the Social Rate of Discount," *Quarterly Journal of Economics*, Vol. 84, No. 3 (1970), pp. 430–450.
4. *Ibid.*
5. These arguments assume full employment.
6. This formulation assumes that labor and capital employed in coal distribution are not perfectly mobile.
7. In the early sixties (1960 to 1962), the ratio between proved reserves and net production was about 20. This ratio dropped to about 18 in 1965 and to 13 in 1971.
8. Gas reserves are "made up of two broad categories: (1) New Discoveries, ND, which represent the amount of recoverable gas estimated to exist in newly discovered reservoirs; (2) Extension and Revisions, XR, which consist of additions to (or subtractions from) the initial

estimates of gas discovery, due to changing economic conditions or the availability of new information on reservoir size or reservoir characteristics (such as permeability, porosity, interstitial water, and so forth)." See D.J. Khazzoom, "The FPC Staff's Econometric Model of Natural Gas Supply in the United States, "*The Bell Journal of Economics and Management Science*, Vol. 2, No. 1 (1971), p. 53.

9. E.W. Erickson and R.M. Spann, "Price, Regulation, and the Supply of Natural Gas in the United States," in *Regulation of the Natural Gas Production Industry*, ed., K.C. Brown (Baltimore: Johns Hopkins Press, 1972), pp. 192–218.
10. Khazzoom, *op. cit.*, pp. 51–93.
11. American Gas Association, Department of Statistics, *Gas Facts* (Arlington, Va., 1972).
12. Khazzoom, *op. cit.*, p. 75.
13. Peoples Gas, Light, and Coke Company, *Annual Report, 1971* (Chicago, 1972).
14. Future Requirements Agency, *Future Gas Requirements of the United States*, Vol. 4 (Denver: Denver Research Institute, University of Denver, 1971), p. 12.
15. The 1973 present values of cost per unit of gas are about the same assuming a 10 percent discount rate—that is,

$$1.50 \sum_{t=0}^{17} \left(\frac{1.00}{1.10}\right)^t \cong 1.185 \sum_{t=0}^{17} \left(\frac{1.04}{1.10}\right)^t$$

When a 20 percent discount rate is assumed, the present value of an immediate increase of 50 percent is higher—that is,

$$1.50 \sum_{t=0}^{17} \left(\frac{1.00}{1.20}\right)^t > 1.185 \sum_{t=0}^{17} \left(\frac{1.04}{1.20}\right)^t$$

16. All projections assume negative exponential decay and linear growth. Other projection methods—a transition probability model and non-linear growth—were also tried. Each of these methods had shortcomings; the one used had the least deficiencies.
17. A description of the computer model used to estimate conversion and space heating costs is provided in Appendix C. The relationships between rents, conversions, and abandonments are discussed in Appendix D.
18. See Appendix D for the analysis used to obtain the cost figures.
19. Personal conversations with the Department of Environmental Control's personnel.
20. See Appendix G for more details on endogenous and exogenous price changes.

Chapter Three

Air Quality Impacts

INTRODUCTION

There are two viewpoints associated with the abatement of air pollution. The first is the global approach that emphasizes the total quantity of pollutants emitted into the atmosphere. Advocates of this view do not consider the distribution of these pollutants in the atmosphere. The second viewpoint is the air quality approach that confines itself to the levels of air pollution on or near the ground–i.e., the pollution that directly affects humans, animals, vegetation, and materials. Considerations such as the spatial patterns of emission sources, the height of emission releases, and meteorological conditions are of primary importance in this approach.

The federal government in establishing air quality standards has explicitly taken the second viewpoint. Since there are no data reflecting the adverse effects of pollution in the upper atmosphere and since there are data and techniques available for determining the damage effects of air pollution at ground level, we also take the air quality approach.

Because there is more than one way to improve air quality, it is necessary to select the best policy based on a given decision criterion. The decision criterion proposed by the federal government is to minimize the cost of attaining specific air quality standards by 1975 and maintaining them in the future. The decision criterion advocated and used in this study is to maximize the present value of net social benefits (benefits minus costs) of pollution control policies, subject to an equitable distribution of their impacts.

A highly sophisticated procedure is employed to estimate the emission reductions, over a given planning horizon, resulting from a control policy, and to translate these reductions into improvements in air quality. These air quality estimates, coupled with the cost estimates made in the previous section, can be used to make a decision based on the federal criteria. Similarly, the improvements in air quality can be used as input into a benefit calculation to determine the most beneficial control policy for residential space heating sources. Benefit evaluation is discussed in detail in the next chapter. The remainder of this chapter will be devoted to describing the analytical procedure used to estimate the effectiveness of each control policy and to presenting the results of the analysis.

ANALYTICAL PROCEDURE

The analytical system employed is a synthesis of many models. It is utilized to test and evaluate the effectiveness of various emission control strategies. A schematic of this system is shown in Figure 3–1.

Emission Inventory

The most important set of data for environmental planning is the emission inventory. In addition to emission rates, it provides information on location, fuel consumption, capacity, and stack characteristics of sources. They are also typed by their Standard Industrial Classification (SIC) code and by emission source type—i.e., as either fuel combustion, process or incinerator sources. Fuel combustion sources are those that emit pollutants from the direct combustion of fossil fuels; process sources emit pollutants from manufacturing activities other than fuel combustion; and incinerator sources emit pollutants from the burning of solid wastes. Table 3–1 lists the data provided in the emission inventory of the Chicago Metropolitan Air Quality Control Region (CMAQCR). The CMAQCR is comprised of 8 counties: DuPage, Cook, Kane, Lake, McHenry, and Will in Illinois, and Lake and Porter in Indiana.

The large emitters of sulfur dioxide or particulate matter are included

Table 3–1. Emission Inventory Parameters

Source Characteristics:	Emissions:
Source I.D. Number	Sulfur Dioxide
SIC Code	Particulates
Political Jurisdiction	Fuel Combustion Source Characteristics:
X Coordinate	Heat Value of Coal
Y Coordinate	Heat Value of Oil
Source Type, i.e.,	Heat Value of Natural Gas
Fuel Combustion	Coal Consumption
Process	Oil Consumption
Incinerator	Gas Consumption
Land Area (for area sources only)	Process Source Characteristics:
Stack Characteristics:	Process Weight Rate
Exhaust Gas Volume	Incinerator Source Characteristics:
Exhaust Gas Temperature	Process Weight Rate
Exhaust Gas Exit Velocity	
Stack Height	
Stack Diameter	

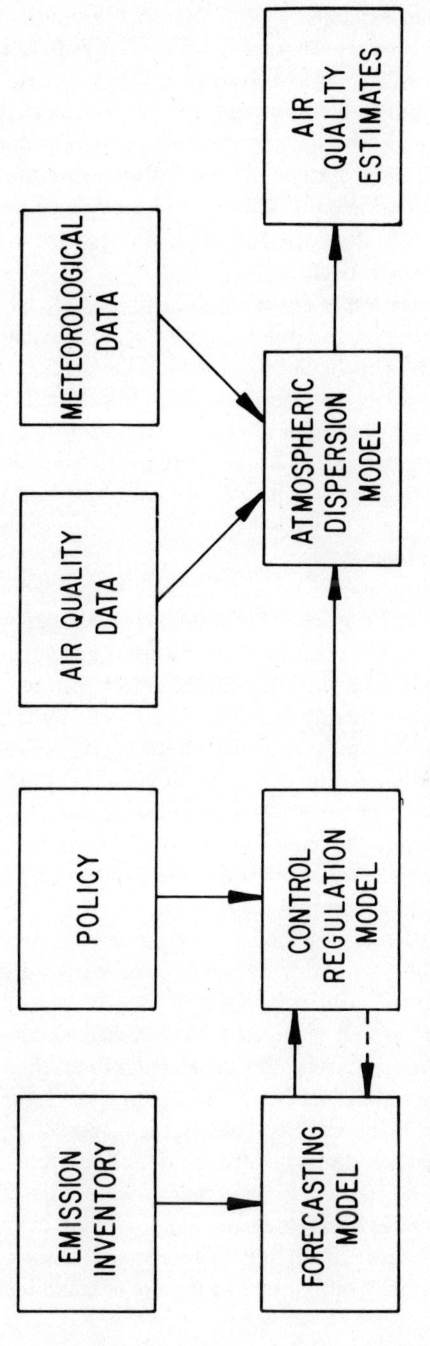

Figure 3–1. Air Pollution Control Model.

as individual sources in the emission file. These are called point sources and may even represent a single stack of a firm. The smaller sources are aggregated geographically into area sources. These area sources range in size from one square mile in the city of Chicago to 100 square miles in the surrounding rural areas. The SIC code is used to identify point sources as residential, commercial, utility, or industrial sources. Separate residential, commercial, and industrial area source emission files were generated. The emission inventory used in the analysis has 459 sulfur dioxide point sources, 456 particulate point sources, and 324 area sources within the CMAQCR. The sources were inventoried in 1968, 1970, and 1968, respectively.

The disaggregation of entries in the emission file by economic activity, emission source type, and political jurisdiction provides considerable flexibility in the analysis. This flexibility is used to great advantage in the forecasting and regulation models that are discussed below. It is through these models that the emission inventory is altered to account for economic growth and pollution control regulations. In some instances the control policy may affect projected economic growth, which further affects emissions.

Forecasting Model

The forecasting model projects emissions from the utility, industrial, commercial, and residential sectors of the economy. Although we were primarily concerned with the changes in residential emissions, the "background" generated by the utility, industrial and commercial sources had to be calculated to obtain estimates of total air quality to compare with standards.

Industrial emissions were projected on a SIC basis and a constant growth rate for industrial area sources was employed. This procedure assumes growth will occur in place–i.e., at existing locations of industrial activity. The growth rates employed were obtained from the U.S. Department of Commerce publication *Economic Projections for Air Quality Control Regions*.[1] The utility projections were based on fuel usage, fuel conversion plans, boiler ages, plant capacities, and retirement and construction schedules for each power station in the region. These data were obtained from written questionnaires and personal interviews with the utilities involved.[2]

The most important projections for our analysis are the residential emission forecasts. These forecasts were considered a function of the space heating regulation being simulated. The control policy was assumed to affect emissions in two ways: (1) by causing a shift in the market shares of each fuel; and (2) by limiting emissions through sulfur content limitations. The market shares of each fuel may be affected in a number of ways. First, the regulation may cause the relative prices of the fuels to change making conversions more attractive. This is what happened when the Chicago sulfur law was passed. Secondly, the regulation may outlaw or ban the use of some fuels forcing conversions to other fuels. This would be the situation with the coal or the coal and oil ban.

The forecasting model simulates these effects by using Peoples Gas' household inventory data and the conversion algorithm described in Chapter 2. The output is an estimated future inventory of residential dwelling units heated with coal, oil, gas, and electricity in Chicago. The inventory is disaggregated by building size and square mile areas that correspond to the area sources in the emission inventory. A different inventory is projected for each control policy reflecting the fuel market adjustments caused by each policy.

Recall again from Chapter 2 that fuel consumption figures by building size class were multiplied by the number of dwelling units in each class to obtain fuel consumption estimates. Let H_{ik}^{p} be the quantity of fuel i consumed in area k under policy p. Then emissions can be calculated as

$$E_{jk}^{p} = \sum_{i} (H_{ik}^{p})(EF_{ij}^{p}), \qquad (1)$$

where

E_{jk}^{p} ≡ emissions of pollutant j in area k under policy p;

EF_{ij}^{p} ≡ emissions of pollutant j per unit of fuel i consumed under policy p.

The effect of restricting the sulfur content of each fuel is reflected in the emission factor, EF, employed. The sulfur contents assumed for the no control case are 3 percent for coal, 2.5 percent for residual oil, and 0.3 percent for distillate oil. When a sulfur limit is imposed the sulfur content of each of these fuels is less than or equal to the limit. When the coal ban is applied, a sulfur limit of one percent is used for oil. Table 3-2 shows the emission factors used for this analysis. The ash content for coal was set equal to 10 percent and was not affected by the control policies modeled. The particle size factors employed were: coal .5, oil 1.0, and gas 1.0. These factors reflect emissions of particles less than 100 microns in size.[3]

The data base for the commercial activity was not as comprehensive or as reliable as the residential data. The commercial data consisted of emissions of sulfur dioxide from commercial sources for each area source in the emission inventory and the square footage of commercial activity using each fuel for space heating. Market shares for commercial fuel consumption were calculated using the square footage data. Then the number of dwelling units in each square mile using each fuel was estimated such that: (1) they would emit the amount of sulfur dioxide attributable to them in the emission inventory,

Table 3-2. Emission Factors for Residential Space Heating

	Coal (lbs/ton)	*Oil ($lbs/10^3$ gal)*		*Gas ($lbs/10^6 ft^3$)*
		Residual	Distillate	
Particulates	(2)(A)P	(22.6)P	(10)P	(19)P
Sulfur Dioxide	(38)S	(156.6)S	(142)S	0.6

Note: A is the ash content of the fuel.
P is the particle size factor for the fuel.
S is the sulfur content of the fuel.

The residual figures are 95 percent of the residual plus 5 percent of the distillate figures listed under "industrial and commercial"–i.e., they are weighted averages of U.S. EPA emission factors. The distillate figure is listed under "domestic" by the EPA.

Source: U.S. Environmental Protection Agency, *Compilation of Air Pollutant Emission Factors,* (Second Edition), Research Triangle Park, North Carolina (1973), pp. 1.1–3, 1.3–2, and 1.4–2.

and (2) their estimated fuel consumption would be consistent with the calculated fuel market shares for commercial establishments. The resulting number is referred to as the number of equivalent dwelling units.

The equivalent dwelling units were then used to calculate the sulfur dioxide and particulate emissions as a function of each space heating control policy as was done for the residential sector. The only difference is that commercial activities were held constant at their 1968 levels—that is, growth in commercial activity was not considered and emissions changed only as a result of the control regulation. This assumption hardly affects the air quality estimates since most new commercial buildings are heated by natural gas or electricity, both of which are clean energy sources.

Because of the poorer quality of the commercial data, the cost and benefit calculations were conducted only for the policies as they apply to residential activities. The commercial calculations were used only to estimate 1975 air quality levels and to compare them with standards.

Control Regulation Model

The control regulation model simulates the effects of various controls on emissions. Separate regulations for fuel combustion, process and incinerator sources are applied to utility, industrial, commercial, or residential activities. Different regulations are also applied to existing and new sources in the CMAQCR. The utility and industrial regulations assumed for this study are those promulgated as part of Illinois' and Indiana's air pollution implementation plans and the city of Chicago's low sulfur ordinance. These regulations are summarized in Appendix A. The space heating controls are the same as the utility and industrial fuel combustion regulations. In Illinois the residential fuel use assumptions are varied for each control policy as described in the section on the forecasting model.

Atmospheric Dispersion Model

The preceding two sections describe the methods used to convert a base year emission inventory into estimates of future emissions reflecting both the economic changes within the region and air quality control regulations. For any future year the forecasted emission inventory is used as input into an atmospheric dispersion model.

The dispersion model converts emissions measured on a mass-per-unit-time basis (e.g., tons per hour), to air quality measured on a mass-per-unit-volume (e.g., micrograms per cubic meter) basis. The true relationship between air pollution emissions and air quality is quite complicated and involves, among other factors, laws of physics, chemistry, aerodynamics, and thermodynamics.[4] Because the physical and chemical relationships affecting dispersion are not completely understood, dispersion models are developed from experiments designed to characterize in quantitative terms the effects of various parameters on an emission plume.

Many multiple source atmospheric dispersion models explicitly relate air quality to emissions, source (stack) characteristics, meteorological conditions, spatial patterns, and time. The emission inventory described previously provides the needed emission and source characteristic data for these models.

For any point source in the region an idealized picture like that shown in Figure 3-2 can be drawn. In this picture, the wind is blowing in the *X* direction, the *Y*-axis is horizontal in the crosswind direction, and the *Z*-axis is vertical. The effective stack height (or effective height of emission release) is the height at which the plume center line becomes horizontal. The effective stack height is the sum of the physical stack height and an incremental factor related

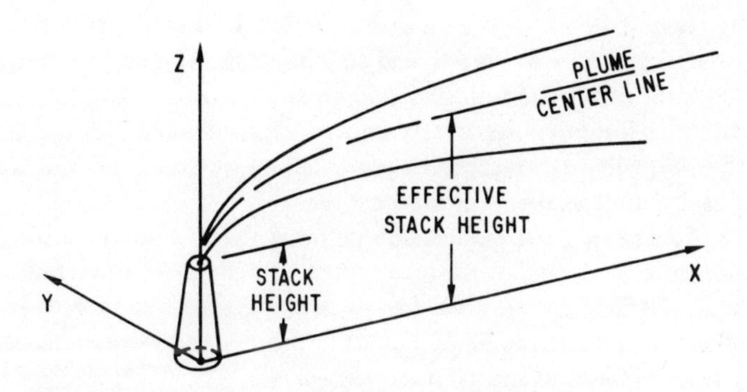

Figure 3-2. Emission Plume from a Single Stack.

to the buoyancy and vertical momentum of the effluent. This incremental factor is known as the plume rise. Many studies have been conducted to estimate plume rises.[5] In general, plume rise is considered a function of the stack diameter; exit gas velocity, volume and temperature; atmospheric temperature and pressure; wind speed; and stability class. The stability of the atmosphere is a measure of how slowly a plume from a stack will grow in cross-sectional areas and is related to the turbulence in the atmosphere.

The transport of the plume in the downwind or *X*-direction primarily results from the wind pushing the plume. As the plume travels downwind, it also disperses horizontally and vertically. The horizontal and vertical dispersion is a function of the stability of the atmosphere and the distance traveled away from the source. Other factors affecting the dispersion in the vertical direction include gravity for particulate matter and the depth of the mixing layer, within which pollutants are confined. (The earth's surface and the mixing layer of the atmosphere act similarly to the walls of a closed container, confining the pollutants between them.) Finally, as the pollutant is transported through the atmosphere, it may decay or react with other elements forming a different substance. The removal or decay of sulfur dioxide is simulated by incorporating its estimated half life–i.e., the time it takes half of any given amount of the pollutant to be removed or converted to other substances in the atmosphere–into the model.

The horizontal and vertical dispersion is often considered to be normally distributed about the centerline of the plume. Models based on this assumption are known as Gaussian dispersion models. Experimental results providing estimates of the horizontal and vertical dispersion coefficients based on the normal assumption have been provided by Pasquill and Gifford.[6] The dispersion model used in this analysis is a modified version of the Air Quality Display Model and is based on a Gaussian diffusion equation.[7] It is a multiple point and area source model that computes annual mean ground level pollutant concentrations at up to 237 locations referred to as receptors. The model calculates the effects of each source on each receptor for 480 observed combinations of wind direction, wind speed, and stability class. The relative frequency of occurrence for each meteorological combination is used to weight air quality estimates computed separately under each combination of conditions. Finally, the air quality at a receptor is estimated by summing the weighted effects of each source at the receptor location.

Area sources are modeled by defining a virtual point source–i.e., a hypothetical source upwind of the area source that simulates the effects of a uniform emission from a large area. The stack heights for area sources (virtual point sources) are assumed to be 40 meters in the central business district and industrial areas of Chicago and 10 meters elsewhere.

Input data to this dispersion model consists of ambient air quality data, an emission inventory, and meteorological information. An important feature of AQDM is that it is a calibrated model. Calibration is accomplished by using the model to estimate air quality at locations where actual air quality is measured, regressing these estimates against the observed values, and adjusting air quality estimates at other locations so they are consistent with observed air quality in the region. The regression curves used for the sulfur dioxide and particulate calibration are $Y = 27.2 + 0.29X$ and $Y = 58.0 + 0.57X$, respectively, where Y is the observed and X is the calculated air quality. The coefficients of determination, R^2, are .254 and .464, respectively. The standard errors of the X-coefficient estimates are .095 and .069, respectively. In the Chicago region, data from 82 particulate and 30 sulfur dioxide air quality monitors were used to calibrate the model.

The model was run twice (because of the 237-receptor point limitation of the model) to obtain air quality estimates at 452 separate locations. Air quality estimates were made on a square mile basis in the city of Chicago and at receptors 4 miles apart in the rest of the region.

Air Quality Estimates

The outputs of the analytical procedure just described are estimates of air quality for any future year for a given control policy. These estimates are displayed in tabular form, as isopleth maps and probability contours (isoprobs). The tabular data are used as inputs for the benefit calculations discussed in Chapter 4. Isopleth maps show contours of equal air quality. This means that at any point on a contour line the air quality is equal to the value of that contour. The air quality between two contour lines is somewhere between the value of the two contours. Therefore, the higher the value of the contour the more heavily polluted the area. The spread between contours indicates the extent of the pollution and the rate at which it improves or worsens with distance. Referring to Figure 3-3, this means that in the center portion of the city of Chicago the sulfur dioxide levels in 1970 were greater than 100 $\mu g/m^3$. Sulfur dioxide levels improve rather quickly and most of the Illinois portion of the region is below 60 $\mu g/m^3$ (the secondary standard for sulfur dioxide before it was abolished). The primary standard of 80 is violated in most of the city of Chicago and the Gary-Hammond area. As can be seen, isopleth maps are useful for indicating the expected air quality concentrations over the entire region and for comparing these estimates with federal standards.

However, atmospheric dispersion models do not completely describe the dispersion process as indicated by the relatively poor fit of the calibration curve. Therefore, air quality projections are most appropriately expressed in probabilistic terms. By using the statistical nature of the regression procedure used to calibrate the model, it is possible to quantify some of the errors associated with the model. Figure 3-4 shows the sulfur dioxide calibration curve. Some of the data points are shown to emphasize that the regression curve represents only the center of a statistical distribution of these data. In addition, confi-

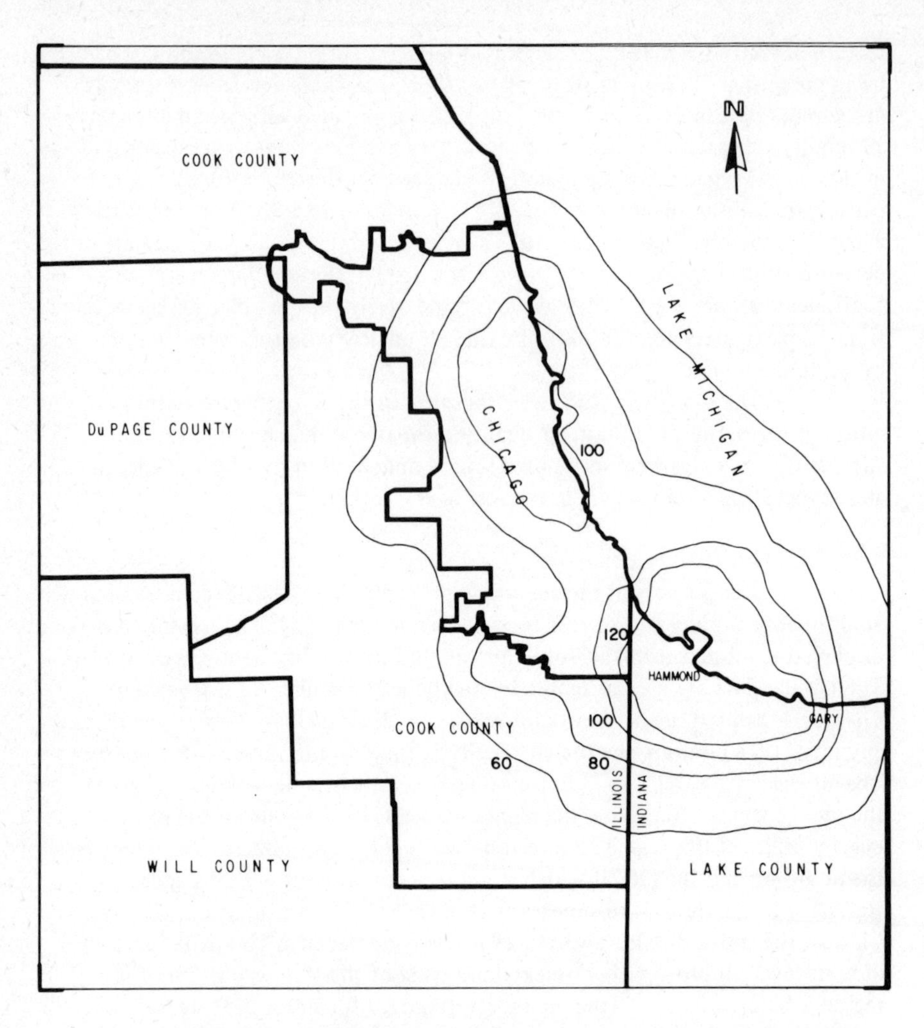

Figure 3-3. Estimated Sulfur Dioxide Air Quality Levels in 1970 (Annual Arithmetic Means, $\mu g/m^3$).

dence bands about the curve have been drawn. Let these bands represent 80 percent conficence. Then when the model calculates a value of C, and if the emission patterns are correctly modeled, the probability that the actual air quality will be between P and P' is .8–i.e., it will tend to be within this range 8 out of 10 times. One out of ten times, the actual value will tend to be greater than P and one out ten times it will tend to be less than P'. Similarly, if the model calculates C', then the probability that the actual air quality will be below 80 is .1 and the probability that it is greater than 80–i.e., that a violation of the sulfur dioxide standard will occur–is .9.

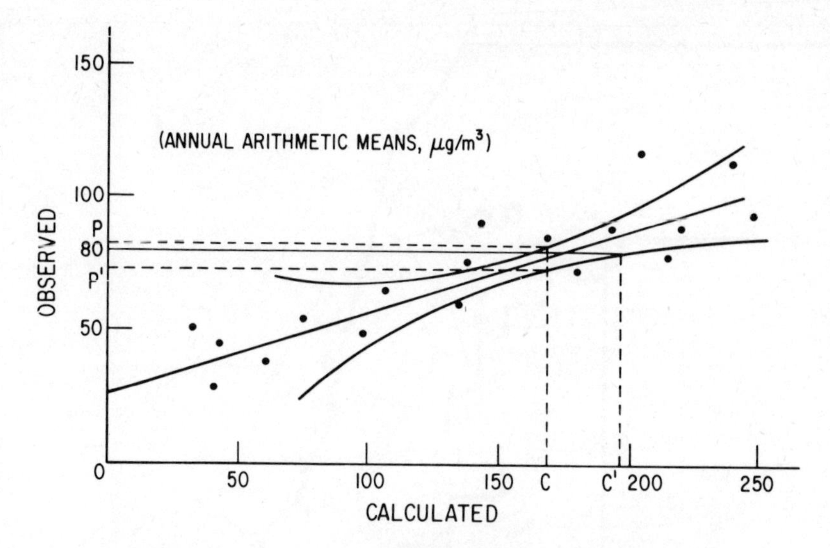

Figure 3-4. Sulfur Dioxide Curve with Confidence Bands.

From the isoprob map shown in Figure 3-5, we see that the probability of a violation of the sulfur dioxide standard in the central portion of the city of Chicago in 1970 is .7, and the probability of a violation in Indiana is .9. Recall from Figure 3-3 that indeed this entire area was estimated to be in violation of the primary federal standard of 80 $\mu g/m^3$. (Note that the isoprob showing a probability of .5 is equal to the isopleth for 80 $\mu g/m^3$. This occurs because the isopleth is an expected value and on the average the actual air quality will be above the expected value 50 percent of the time.) Actual measurements of sulfur dioxide levels in 1970 were indeed above the standard in the central portion of Chicago and the Gary-Hammond area.

An important issue related to the isoprobs is the difference in viewpoint between the confidence of violating standards and the confidence of meeting standards. If the position is taken that the standard must be met at any cost, then the probability of meeting the standard should be maximized (minimize the probability of violation) subject only to technical constraints. On the other hand, if the position is taken that a regulation restricts the free market and should only be enacted if absolutely necessary, then the probability of violating the standard should be very high before deciding that a regulation is necessary.

RESULTS OF AIR QUALITY ANALYSIS

Estimates of total air quality levels in 1975 were generated for the no control, the one percent sulfur, the 1975 coal ban, and the 1975 coal and oil ban space

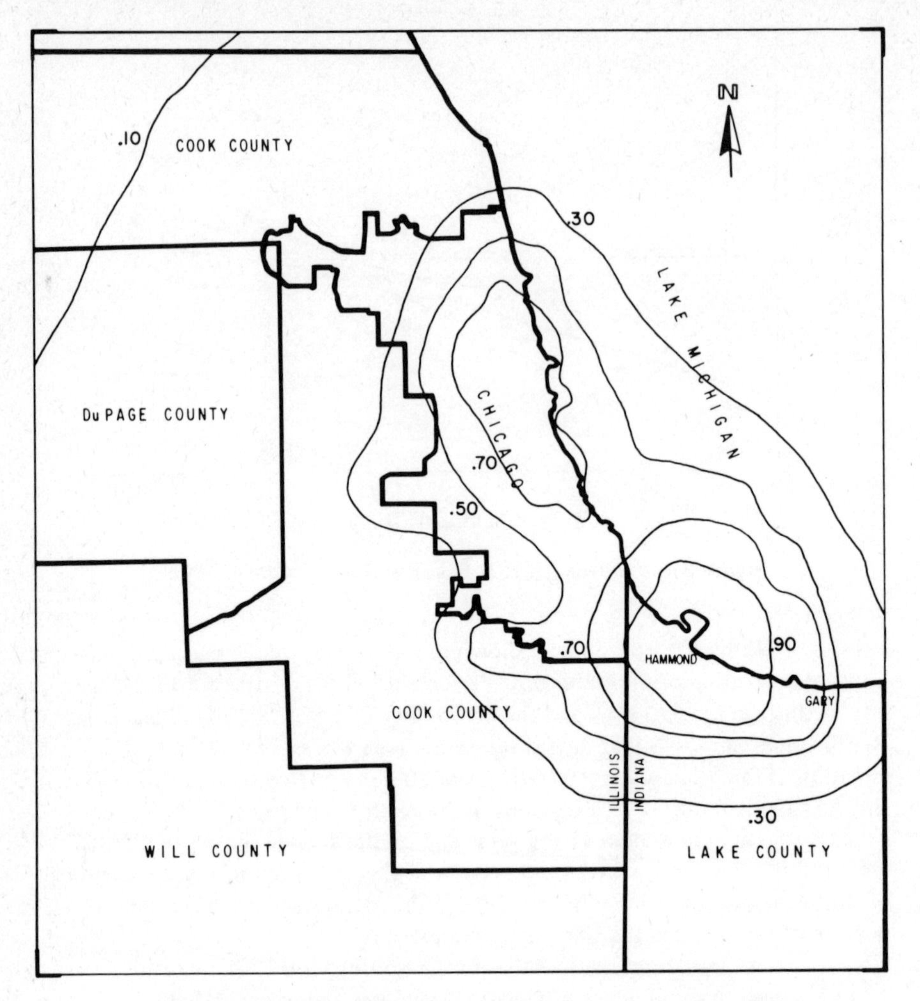

Figure 3-5. Probability Contours for Violations of the Primary Sulfur Dioxide Air Quality Standards in 1970.

heating control policies. These estimates assumed 100 percent compliance with the industrial, utility, and commercial control regulations. The estimates of air quality resulting from residential sources only were generated for each square mile in the city of Chicago from 1968 through 1990 for the no control and 1 percent sulfur cases, and from 1972 through 1990 for the fuel ban policies.

The relative impacts of these policies were used to calculate the relative benefits of each policy over a planning horizon. The estimates of total air quality levels were used to determine the necessity and severity of space heating controls to satisfy federal air quality standards in 1975. The results of the latter analysis are presented below.

The isopleth maps in Figure 3-6 show the expected sulfur dioxide levels in 1975 for each of the four residential policies. Map (a) shows the no control case; map (b), the 1 percent sulfur law; map (c), the 1975 coal ban; and map (d), the 1975 coal and oil ban. The expected sulfur dioxide levels improve as residential controls are tightened. Only under the no control situation is there an expected violation in Illinois of the federal annual ambient air quality standard for sulfur dioxide. Under any one of the control situations—i.e., b through d—the probability of violating the standard is less than .5, while under the no control case it is about .65. This indicates that some level of control is necessary and that the sulfur law is sufficient to meet the federal standard for sulfur dioxide.

The situation for particulate matter is quite different. Although the entire region will be expected to be in compliance with the primary standard of 75 $\mu g/m^3$, the federal secondary standard of 60 $\mu g/m^3$ is expected to be violated under any one of the space heating control policies. This result is shown in Figure 3-7. The probabilities of violating the federal secondary standard in 1975 are shown in Figure 3-8. These figures at first are deceiving since they may imply significant reduction in the probability of a violation as space heating controls are tightened. Actually, the maximum probability of a violation under the no control case is .75; under the 1 percent sulfur law, .72; under the coal ban, .69; and under the coal and oil ban, it is still .68. These results seem to indicate that there is very little to be gained by a strict space heating control policy, but it might be required in order to approach the standards.

If we assume that the industrial and utility controls are inflexible, then there are three approaches a decision-maker could take on the basis of the information provided: (1) the maximum level of control might be adopted—i.e., the coal and oil ban—based on the argument that the best available control technology must be applied if standards are not reached by 1975, and the fact that there is a high probability of a violation of the secondary federal particulate standards in 1975; (2) the minimum level of control might be adopted—i.e., no control—based on the argument that there is not a significant improvement in air quality to justify stricter controls; and (3) an arbitrary cutoff point for the probability of a violation might be specified. If the desired probability is greater than .75, then no residential controls will be imposed; if it is less than .68, then the coal and oil ban will be selected. Any probability between these two extremes would result in the sulfur law or the coal ban being selected.

Therefore, a decision on the level of control for space heating sources can be made with the information generated, provided a decision criterion is specified. Since there are at least the three reasonable decision criteria discussed above and each produces a different solution, additional information would be desirable to reduce some of the subjectivity in the decision process. This additional information includes the costs discussed in the previous chapter and the benefit calculation discussed in the next chapter.[8]

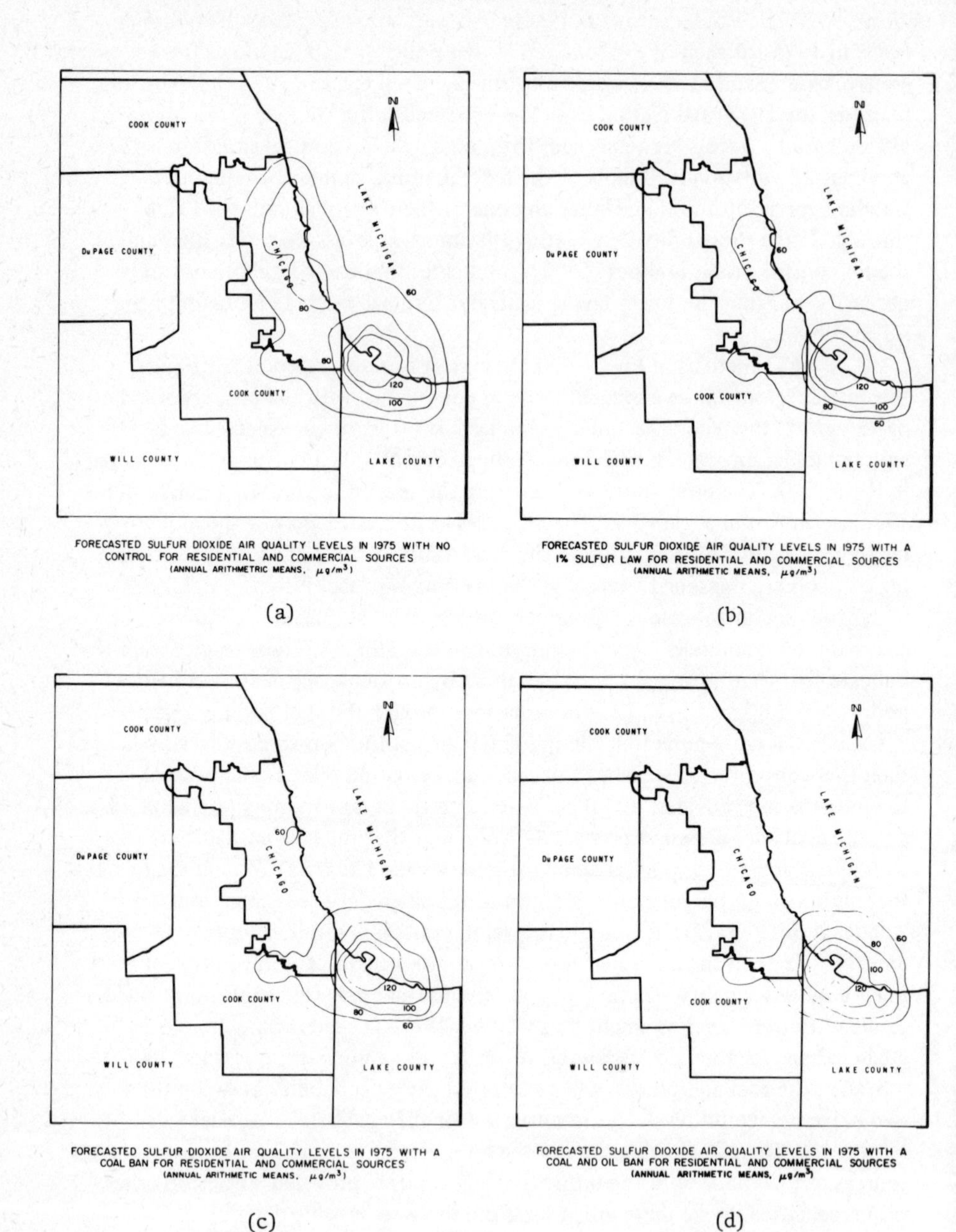

FORECASTED SULFUR DIOXIDE AIR QUALITY LEVELS IN 1975 WITH NO CONTROL FOR RESIDENTIAL AND COMMERCIAL SOURCES (ANNUAL ARITHMETIC MEANS, $\mu g/m^3$)

(a)

FORECASTED SULFUR DIOXIDE AIR QUALITY LEVELS IN 1975 WITH A 1% SULFUR LAW FOR RESIDENTIAL AND COMMERCIAL SOURCES (ANNUAL ARITHMETIC MEANS, $\mu g/m^3$)

(b)

FORECASTED SULFUR DIOXIDE AIR QUALITY LEVELS IN 1975 WITH A COAL BAN FOR RESIDENTIAL AND COMMERCIAL SOURCES (ANNUAL ARITHMETIC MEANS, $\mu g/m^3$)

(c)

FORECASTED SULFUR DIOXIDE AIR QUALITY LEVELS IN 1975 WITH A COAL AND OIL BAN FOR RESIDENTIAL AND COMMERCIAL SOURCES (ANNUAL ARITHMETIC MEANS, $\mu g/m^3$)

(d)

Figure 3–6. Sulfur Dioxide Air Quality Estimates for 1975.

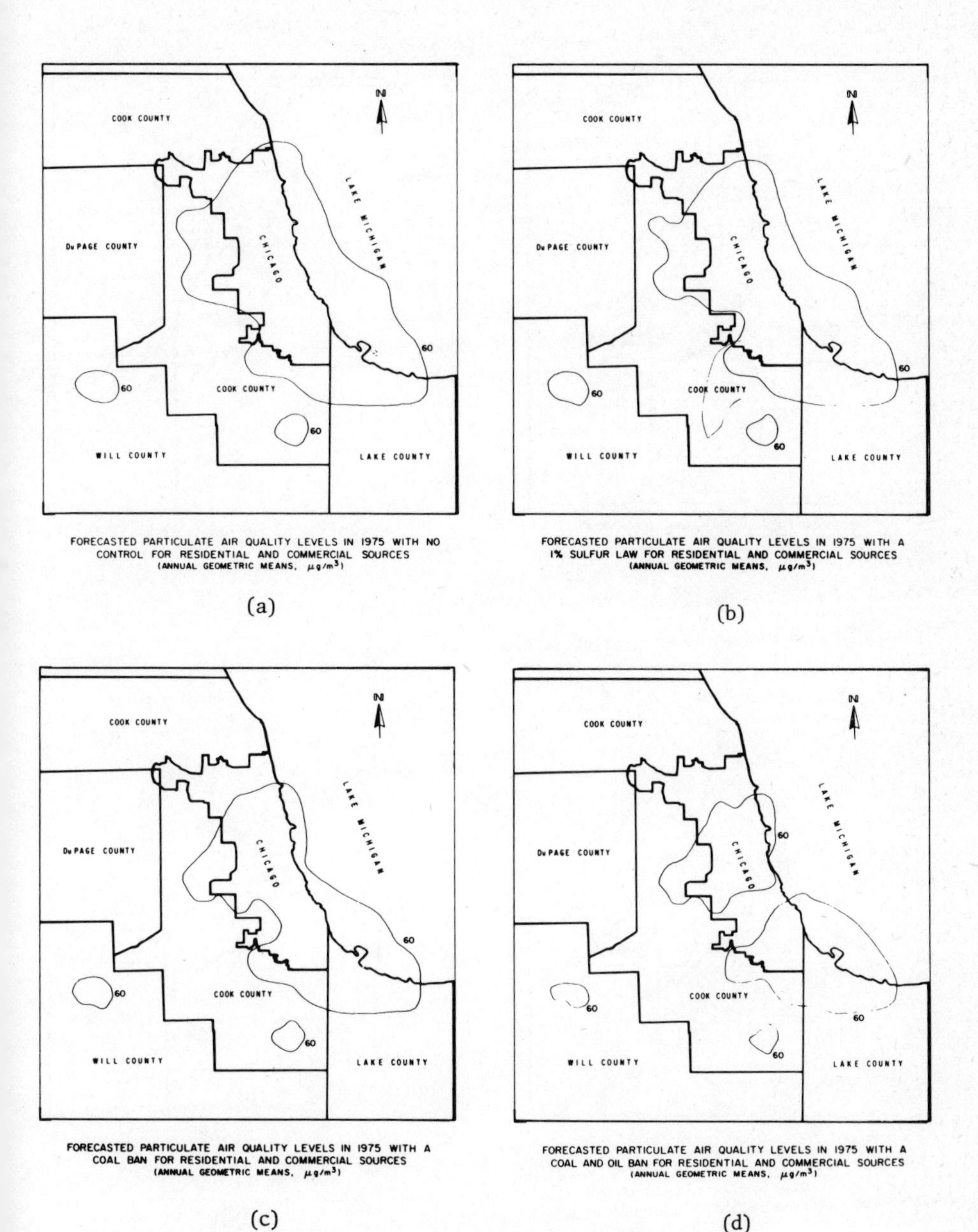

Figure 3-7. Particulate Air Quality Estimates for 1975.

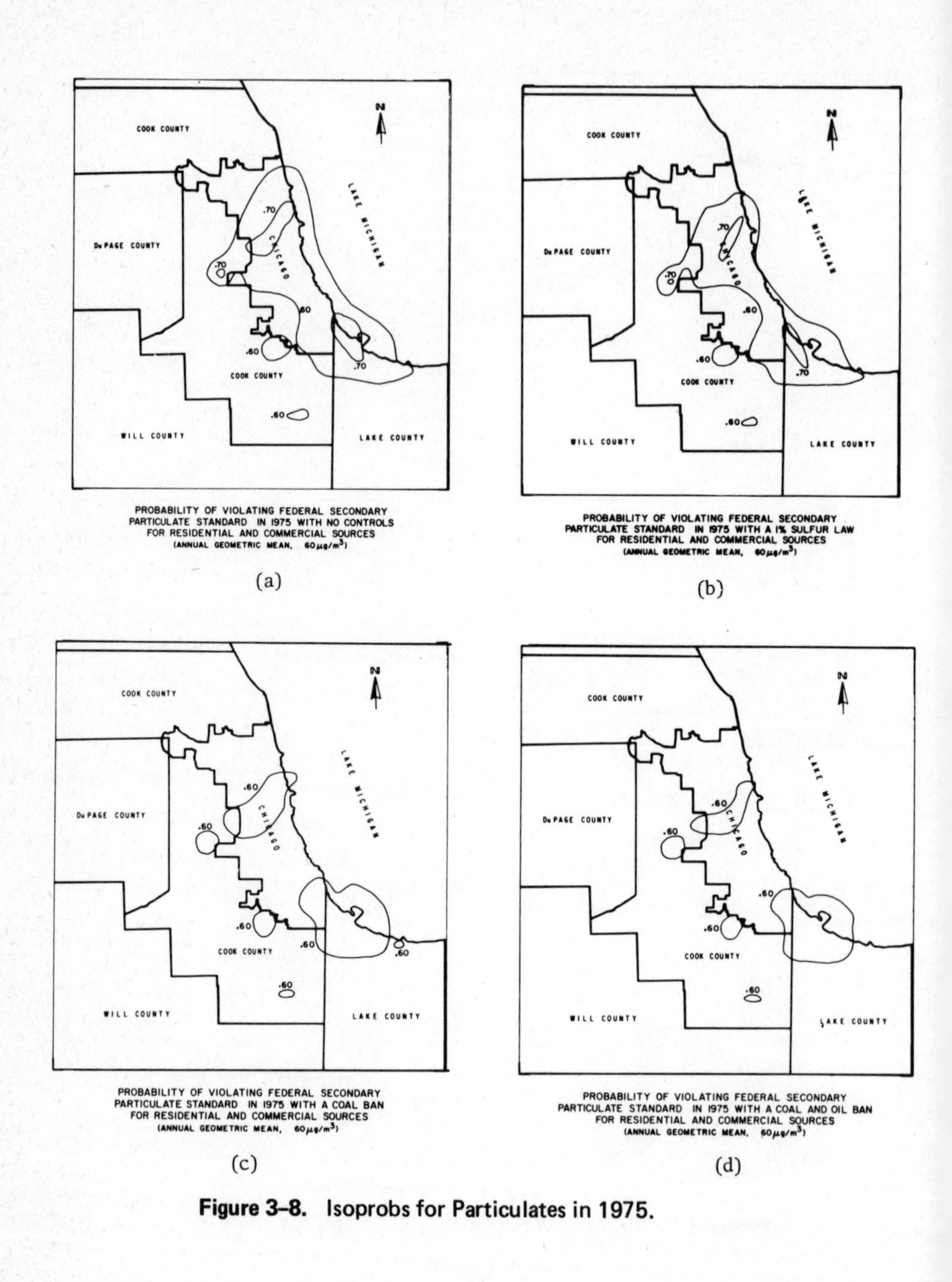

Figure 3–8. Isoprobs for Particulates in 1975.

NOTES TO CHAPTER THREE

1. U.S. Department of Commerce, Office of Business Economics, *Economic Projections for Air Quality Control Regions* (Washington: U.S. Government Printing Office, 1970).
2. For more details on the industrial and utility forecasting technique, see A.S. Cohen, et. al., *Growth Analysis Report for the Illinois Implementation Planning Program* (Argonne, Illinois: Argonne National Laboratory, Center for Environmental Studies, 1972).
3. U.S. National Air Pollution Control Administration, *Air Quality Criteria for Particulate Matter*, Publication No. AP–49 (Washington: U.S. Government Printing Office, 1969), p. 27
4. For a detailed description on atmospheric dispersion, see D.B. Turner, *Workbook of Atmospheric Dispersion Estimates* (Cincinnati: Air Resources Field Research Office, Environmental Science Service Administration, 1970).
5. See, for example, G.A. Briggs, *Plume Rise* (Oak Ridge: Oak Ridge National Laboratory, U.S. Atomic Energy Commission, TID–25075, 1969); J.R. Holland, *A Meteorological Survey of the Oak Ridge Area*, Report ORD–99 (Washington: U.S. Atomic Energy Commission, 1953); and J.E. Carson and H. Moses, "The Validity of Currently Popular Plume Rise Formulas," in *Proceedings USAEC Meteorological Information Meeting* (Clark River, Ontario, 1967), pp. 1–20.
6. F. Pasquill, "The Estimation of Dispersing of Windblown Material," *The Meteorological Magazine*, Vol. 90, No. 1063 (1961), pp. 33–49; and F.A. Gifford, "Uses of Routine Meteorological Observation for Estimating Atmospheric Dispersion," *Nuclear Safety*, Vol. 2, No. 4 (1961), pp. 47–51.
7. TRW Systems Group, *Air Quality Display Model* (Washington, D.C., 1969).
8. A cost-effectiveness calculation can be made using the cost data from Chapter 2 and the air quality improvement estimates in this chapter. This would provide very little information for two reasons. First, the effectiveness of each regulation is not the same and, therefore, there is no basis for comparison. Secondly, there is no clear cut definition of the effectiveness of a regulation in general. The effectiveness may be defined for a specific receptor within the region or for the average concentration throughout the region. None of these approaches is entirely satisfactory.

Chapter Four

Benefits of Residential Fuel Policies

INTRODUCTION

Willingness to Pay Approach

Among the benefits of air quality improvements are reductions in rates of mortality, the incidence of some diseases, and soiling and deterioration of materials. Most likely an individual would be willing to pay something for these benefits. It is proposed that the aggregate of the maximum dollar values of each person's willingness to pay for an air quality improvement, over all individuals, is an appropriate measure of the social benefit of reducing air pollution throughout a region.[1] This chapter discusses the estimates made to determine how much the people residing in the Chicago area might be willing to pay for the air quality improvements associated with each of the residential fuel policies discussed in the previous chapter. In the next chapter, the willingness to pay estimates are compared with the corresponding social cost estimates presented in Chapter 2.

One of the reasons why people would be willing to pay for air quality improvements is that they increase personal incomes or enable individuals to reduce expenditures for medical care, cleaning, and other related goods and services. Estimates of these gains are made in a number of ways. For persons not yet at retirement age, an extension of life expectancy makes it possible to postpone the date of retirement and earn additional wages or to enjoy additional days of retirement. One way of expressing the value of these added days in dollar terms, regardless of which option the worker actually chooses, is to estimate the potential increase in his earnings assuming retirement is delayed. (For individuals who freely choose retirement rather than additional work, this method underestimates the dollar value of the added days; for persons forced to retire, it may give an overestimate.) An additional effect of reduced mortality rates is a delay in costs of final illness and burial.

The difference between the present value of these costs occurring in one year rather than in a later year is the gain from extending life. Finally, reduced yearly expenditures for health care and material maintenance are estimated. The present value of these savings is a measure of the benefits resulting from air quality improvements.

Each individual should be willing to pay at least the value of these gains for air quality improvements, because he could do so without adversely affecting his level of consumption or expected length of retirement. However, he probably would be willing to pay an even greater amount for air quality improvements. The reasons involve intangibles such as deferring or reducing the pain, anxiety, and suffering of illness or death;[2] extending the time during which the family as a whole can function as a social unit; extending the time over which the individual contributes as a citizen in the society; improving the aesthetic value of the environment; and increasing the enjoyment of outdoor activities. It is assumed that a person would pay for the intangible benefits by foregoing consumption of other goods and services. A highly tentative estimate of the proportion of goods and services an individual would be willing to forego will be made.

One might object to this method of evaluation on the grounds that human life is priceless, of infinite inherent worth, and consequently not capable of being valued in dollar terms. This point of view implicitly underlies the goal of the federal air pollution abatement program, which is to improve the air enough to "protect the public welfare from any known or anticipated adverse effects of air pollution."[3]

To make this objection is to confuse two kinds of normative evaluation processes. Preserving human life is regarded as a "right" in most moral systems; it can appropriately be contrasted only with a "wrong" such as willfully terminating life. In contrast to this kind of evaluation, where there are rules for characterizing an action as right or wrong, benefit–cost comparisons involve a ranking of alternatives.[4] The rhetoric of ranking includes concepts such as "best" and "worst." Selecting the alternative that ranks highest may imply foregoing another judged "right." The rule-oriented approach may at times lead one into inaction because there are no feasible alternatives, or too many. Furthermore, as with environmental policies, the rule-oriented approach may require larger resource investments than would be favored with a ranking approach. These reasons underly this study's advocacy of the benefit-cost ranking approach.

The basis for the benefit-cost ranking is a comparison between the human and material resources expended with each alternative action. In Chapter 2 the term "social costs" was attached to the dollar costs of conversions, fuel consumption, and so forth, to emphasize the fact that these dollar amounts measure the values of the resources required to comply with the policy. Similarly, the estimates of income gains, expenditure reductions, and forgone consumption–i.e., the willingness to pay–can be regarded appropriately as "social benefits." This indicates that they measure the value of resources created as a consequence of the policy, plus the maximum value of resources that individuals would be willing to divert from other uses in order to gain the benefits of the policy. Expressing both the social costs and social benefits in dollar terms provides convenient and comparable measures for both.

A social benefit–cost analysis implicitly makes a ranking of (1) the resource requirements of a given policy, and (2) the other uses that people and institutions would make of the resources in the absence of the policy. Enacting the policy is favored if the aggregate social benefits exceed aggregate social costs. A justification for using this criterion is that when it is satisfied, each person benefited could be charged a dollar amount less than his maximum willingness to pay, and those who incur costs can be fully reimbursed. Therefore, the benefits of the policy can be achieved without adversely affecting anyone.

Of course, such a financing plan is never implemented in practice, since maximum willingness to pay cannot be measured on an individual basis. Therefore, a second analysis in social benefit–cost evaluation must be conducted; namely, identifying those benefited and harmed under the actual financing arrangement, for any policy deemed acceptable by comparison of total social benefits and costs. This is done in the present study. The fuel policies are financed directly by those who must convert to clean fuels, go out of business, or move to a new residence, and by all users of gas. Those benefited are not always required to bear a fair share of the costs. The distribution of benefits and costs is considered in Chapter 5.

Synopsis of Social Benefit Estimates

The effect of each of the fuel policies on air quality has been estimated in each square mile in Chicago and on a four-mile grid in the surrounding area, using the dispersion model described in the preceding chapter. The technique estimates air pollution levels attributable to residential sources for each fuel policy assumption, and compares these estimates for each future year in which the policy is expected to have an effect. For example, Figure 4–1 shows the estimated particulate levels at a location ten miles southeast of downtown Chicago under 6 different policies between 1968 and 1990. (The no control line in Figure 4–1 does not reflect the adjustment for the exogenous coal price increase discussed in Chapter 2.) The maximum impacts of both the 1975 coal ban and the 1975 coal and oil ban are in 1975, when all coal-burning or coal- and oil-burning units should have converted to cleaner fuels. The spatial variation of impacts is illustrated in Figure 4–2. This figure shows the effects of the 1975 coal ban in 1975 in different locations throughout the city. Maps for the other policies, for other dates and for sulfur dioxide would have the same appearance, although the magnitudes of the impacts would differ.

The largest impact in numerical terms is the effect of the 1968 low-

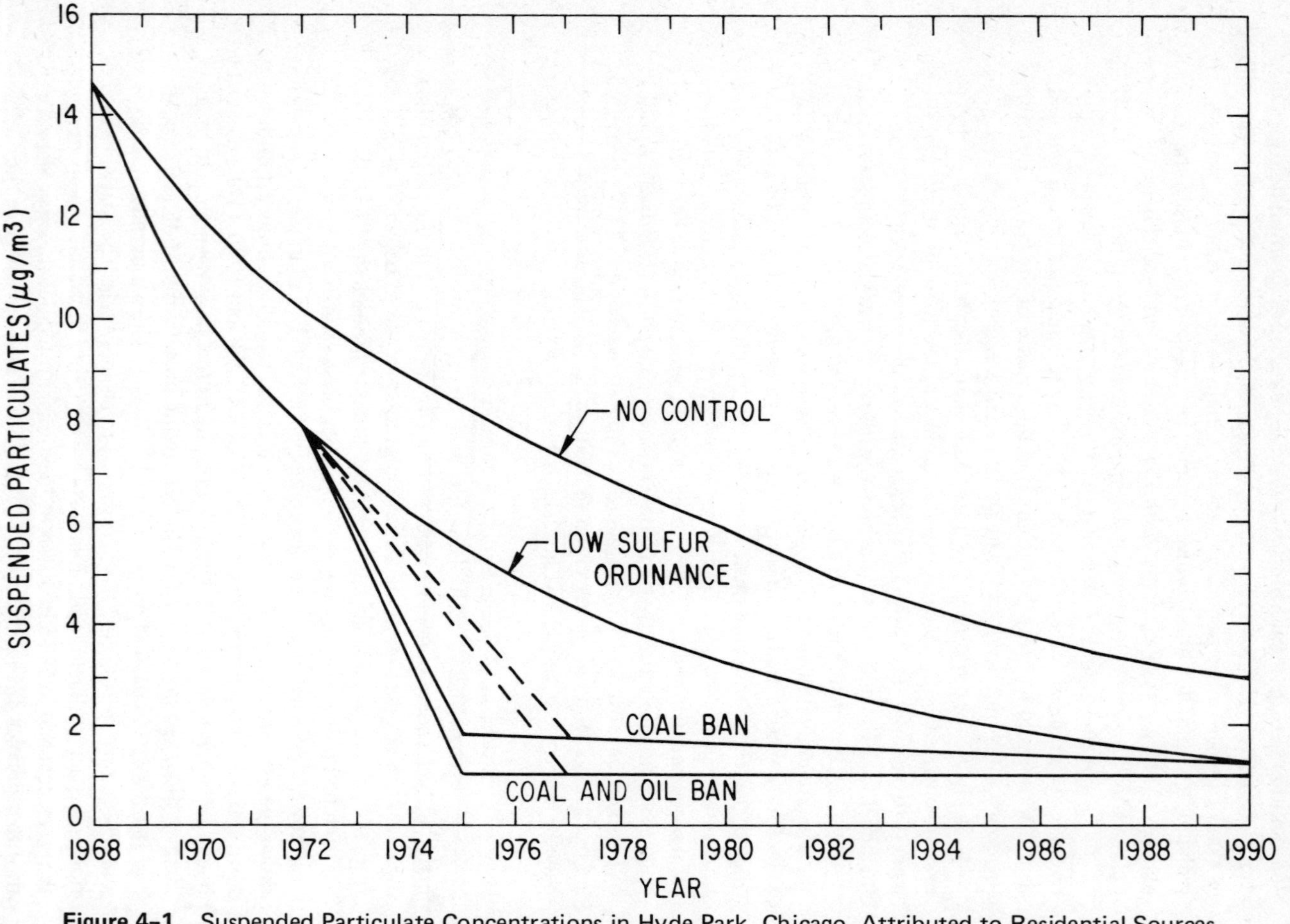

Figure 4-1. Suspended Particulate Concentrations in Hyde Park, Chicago, Attributed to Residential Sources.

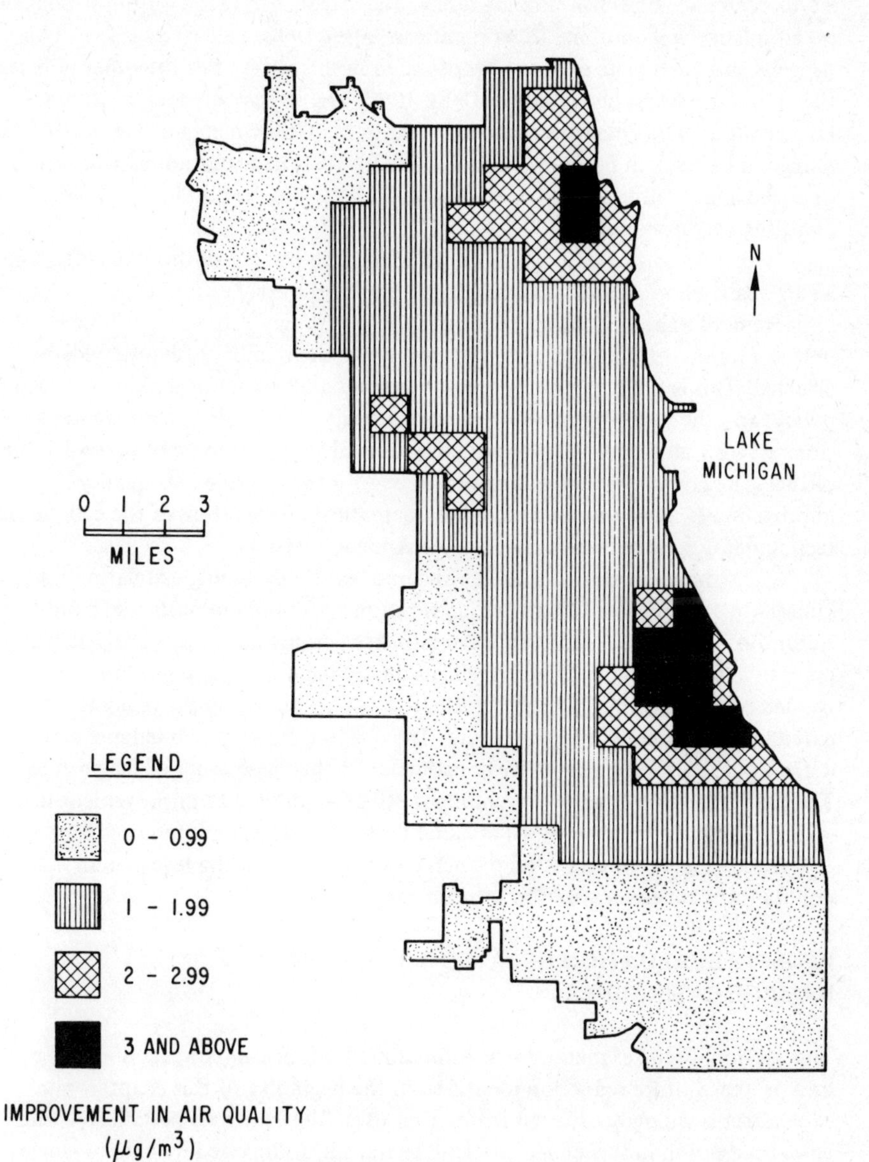

Figure 4–2. Effects of the 1975 Residential Coal Ban on 1975 Average Suspended Particulate Concentrations in Chicago.

sulfur ordinance on sulfur dioxide levels. The effects of the low-sulfur ordinance on air quality will continue to be significant after 1990, the last year for which benefits and costs were explicitly considered in this study. For the other policies, the air quality levels nearly converge by 1990. The largest impacts are in the Hyde Park area of Chicago's south side and in the Uptown area of the north side, indicated in black in Figure 4-2. Both of these areas had substantial proportions of coal-using dwellings in 1968; in this sense, they may be considered as self-polluting neighborhoods.

An estimate of aggregate willingness to pay by all those benefited by an air quality improvement is made by multiplying yearly estimates of the $\mu g/m^3$ improvement in each location by a measure of willingness to pay per household per $\mu g/m^3$ of improvement, and in turn by the number of households affected. This is done separately for particulates and for sulfur dioxide for each policy, and the resulting values are added. Finally, the yearly present values are summed over all future years considered, to provide an estimate of present value of the aggregate maximum willingness to pay for the expected air quality improvements. (For a more detailed mathematical presentation of the estimation techniques described in this chapter, see Appendix H.)

Much of this chapter is concerned with developing estimates of a household's willingness to pay per $\mu g/m^3$ of improvement in particulate and sulfur dioxide levels. A representative estimate coming out of this analysis is a present value of \$36 per $\mu g/m^3$ of sulfur dioxide and \$51 per $\mu g/m^3$ of suspended particulates per household affected, for air quality improvements extending indefinitely into the future. The chapter presents two independent derivations, both of which arrive at estimates of this same order of magnitude. The first derivation considers each major effect of air quality improvement in turn and adds up the results. The second derivation relies on inference from observations on how residential property values vary with the levels of air pollutants between locations in urban areas.

BENEFIT EVALUATION RELATED TO SPECIFIC DAMAGE CATEGORIES

This section discusses methods for estimating the four components of income gain or expenditure reduction identified in the beginning of this chapter, plus expenditures willingly diverted from other uses. The estimates pertain to a one $\mu g/m^3$ reduction in suspended particulate and sulfur dioxide levels, for a single, representative household in the city of Chicago.

Income Effect of Extending the Duration of Life

An analysis of the relationship between air pollution and death rates in urban areas has been conducted by Lave and Seskin.[5] They assembled figures on death rates by age, levels of suspended particulates and sulfates (a chemical constituent of particulates), and on other variables believed related to death rates, including population density, racial composition, and income distribution for 117 metropolitan areas in the United States. Their reported regression coefficients make it possible to construct and examine profiles of probabilities of dying by age under varying particulate levels. [6]

It is assumed that without any improvement in air quality, Chicago age-specific mortality rates in each future year will be the same as the corresponding 1959 to 1961 Chicago average mortality rates. (Lave and Seskin's data are for 1959 to 1961). These mortality rates were adjusted for a one $\mu g/m^3$ improvement in air quality—using the Lave and Seskin coefficients—and a second set of probability profiles for age at death for each present age group was constructed. Table 4-1 shows the expected length of life, and the effect of a reduction in particulates by one $\mu g/m^3$, for four age groups. [7]

The social income gain from increasing a person's expected longevity can be estimated as the difference between the present values of his expected future income with and without the air quality improvement. The major question concerns the appropriate yearly income to be used in estimation of the two present values. Ridker, in an economic analysis of reductions in mortality rates, proposes two possible solutions. [8] One is the additional value of goods and services produced by the individual—i.e., his earnings. The other is the value of goods and services produced less the value of his consumption, to be referred to as net earnings. Ridker points out that the second approach is unsatisfactory

Table 4-1. Effects of a One $\mu g/m^3$ Reduction in Particulate Level on Life Expectancy

Color-Sex Group	*Age, 1972 (Years)*	*Expected Age at Death (Years)*	*Expected Extension of Life (Days)*
White Males	Newborn	65.7	1.9
	35	69.2	1.4
	60	74.5	0.9
	85	88.3	0.4
White Females	Newborn	72.2	2.5
	35	74.7	2.0
	60	78.1	1.5
	85	88.9	0.8
Nonwhite Males	Newborn	60.1	4.3
	35	66.3	3.0
	60	74.7	1.8
	85	90.5	1.0
Nonwhite Females	Newborn	65.1	6.7
	35	69.9	5.3
	60	76.8	3.7
	85	90.7	1.8

because it leads to negative values for retired persons who consume more than they produce. He therefore adopts the first approach; Ridker values all extensions in longevity for people in a given age group by their average earnings and adjusts this figure to account for the probability that some of these people do not participate in the labor force.

The difficulty with Ridker's approach is that it overestimates willingness to pay for persons whose earnings exceed consumption. If these people were to contribute all their expected additional earnings in payment for the environmental improvement, nothing would be left for personal consumption during the extended period of life. For such a person, net earnings are equal to the sum of savings, taxes, and transfer payments to others out of wage income (e.g., support of spouse and children), minus consumption expenditures made with rents or income from capital assets. The terms relating to savings, taxes, and transfers are all gains to others in society if the person lives; the subtracted terms are all monetary losses to others if he lives. Therefore, net earnings are equal to social benefits from a resource point of view and are used in this study to estimate benefits of air quality improvement.

Similarly, gross earnings may also overestimate the willingness to pay for a person whose earnings are less than his consumption. Net earnings for such an individual are negative and are monetary losses to others for as long as he continues living. In such cases, real, intangible benefits to society must exist, which are at least as great as these monetary losses. If this were not true, society would not support these people via intra-family transfers, the social security system, and so forth. In these cases, we conservatively assume that the value of the intangible benefits to others is equal to the amount of the monetary loss, and therefore assign a value of 0 as the income gain component of willingness to pay for extending the expected length of life.

In this study, the net social benefits are approximated under the following assumptions:

1. Earnings exceed consumption expenditures for everyone aged 18 through 64, or until slightly after the 65th birthday, as discussed in numbers 3 and 4 below. Figures on earnings by age in the Chicago area were taken from the *1970 Census of Population*, for age groups 18 through 44 and 45 through 64, for the 4 color-sex groups (see Table 4-1).[9] The figures used are reported average earnings for persons in the labor force multiplied by a labor force participation rate. It was assumed that earnings will grow at a real rate of 2 percent per year, indefinitely into the future.
2. Personal consumption expenditures were assumed to be the same for all age-sex classes within each color grouping. The level of expenditures was derived by dividing the aggregate income of everyone in the Chicago area by the number of persons,[10] and multiplying the result by 0.9 (an assumed proportion of income consumed)[11] for each of the two color groups. It

was assumed that expenditures will grow at a real rate of 2 percent per year, indefinitely into the future. The resulting expenditure levels for 1970 were $3,703 for whites and $2,123 for non-whites. (Figures reported in the census for Negroes were sometimes used for non-whites in this analysis.)

3. The date on which earnings stop, for purposes of estimating earnings gains, is the 65th birthday in the absence of an air quality improvement, or T days after the 65th birthday if there is a one $\mu g/m^3$ particulate reduction, where T is the number of days by which expected longevity at 65 is increased if there is a one $\mu g/m^3$ reduction in particulates.
4. Earnings are 0 for all persons aged less than 18, and for all persons 65 and over (or 65 years + T days if the air quality is improved).[12]

In computing the aggregate earnings gain for the whole population, estimates were made separately for each present age group and multiplied by the number of persons in the age group. Table 4-2 reports figures for the city of Chicago for 4 of the 100 age groups considered, together with aggregate estimates of willingness to pay for the city as a whole.

Deferment of Final Illness and Burial Expenses

Table 4-2 also contains estimates of the gain from deferring the expenses of final illness and burial. They are based on estimates of $1450 for funeral and burial expenses,[13] and $550 for hospitalization and medical treatment during the terminal illness. (This is based on Schrimper's estimate that an average hospital stay for a respiratory condition in Illinois in 1970 was about 7 days and the average cost per day was $81.60. The total cost was $571.[14]) The total expense of $2,000 is assumed to increase at a real rate of 2 percent per year. Because the extension of life expectancy is small, and for most people final expenses occur far in the future, the present value of these benefits is small.

Reduction in Health Care Expenditures

Although evidence on the effect of air quality improvements on the incidence of various diseases is tenuous, it is important to consider the possible magnitudes involved. Health care costs are borne throughout life by most people. To calculate these costs, an estimate of the average reduction in annual health care costs is made corresponding to a one $\mu g/m^3$ improvement in air quality. For each future year the cost reduction is weighted by the probability of a person surviving to that year. (The survival probabilities for each color-age-sex group are derived from mortality rate estimates.) Finally, the present value of these benefits over all future years is calculated.

Lave and Seskin estimated that a 25 percent reduction in the costs of health care for respiratory diseases could be achieved by improving air quality by 50 percent.[15] Assuming that 1968 air quality in Chicago averaged 100 $\mu g/m^3$ for both sulfur dioxide and particulates, a 50 percent improvement would

Table 4–2. Mortality Benefits

		Benefits Per Person (Dollars)		Total, City of Chicago	
Color–Sex Group	Age, 1972 (Years)	Extension of Earnings	Delay of Illness and Burial	Extension of Earnings (Millions of Dollars)	Delay of Illness and Burial (Dollars)
White Males	Newborn	.57	.02		
	35	3.20	.06		
	60	10.62	.12		
	85	0	.12		
	All ages			3.44	74,900
White Females	Newborn	.07	.02		
	35	.41	.06		
	60	2.02	.16		
	85	0	.20		
	All ages			0.57	107,000
Nonwhite Males	Newborn	1.00	.08		
	35	4.18	.18		
	60	10.63	.22		
	85	0	.22		
	All ages			1.56	68,300
Nonwhite Females	Newborn	.28	.10		
	35	1.04	.22		
	60	3.54	.41		
	85	0	.43		
	All ages			0.52	114,000
Total				$6.09 Million	$364,000
Per Household				$5.36	$ .31

Note: The figures are 1972 present values corresponding to a one $\mu g/m^3$ reduction in particulate level. A 10 percent discount rate was used.

be 50 $\mu g/m^3$ for both pollutants. The estimated reduction in costs of treatment for respiratory disease associated with a one $\mu g/m^3$ reduction in both pollutants would then be 0.5 percent of total treatment costs for respiratory diseases (1/50 × .25). It is assumed that the benefits of a one $\mu g/m^3$ improvement in either of these pollutants alone are half of this amount, or 0.25 percent of total treatment costs. Schrimper estimated that $700 million were spent for treatment of respiratory diseases in the Chicago area in 1970.[16] The reduction in respiratory disease treatment costs expected from a one $\mu g/m^3$ improvement in either sulfur dioxide or particulates is therefore $1.75 million ($.0025 \times 700 \times 10^6$) for the Chicago area.

Lave and Seskin also considered treatment costs for other diseases believed to be affected by air pollution.[17] These include cardiovascular disease and cancers other than in the respiratory tract. Their conclusion was that the savings in treatment costs for these diseases would be about two-thirds as much as the savings for treatment of respiratory diseases. Thus, an aggregate estimate of health care cost reductions for the Chicago metropolitan area is about $3 million ($5/3 \times 1.75 \times 10^6$) from a one $\mu g/m^3$ reduction in either of the two pollutants. Since there are about 2 million households in the Chicago area, this is equivalent to $1.50 per household per year. This benefit figure is used to calculate the aggregate present value of health care savings for the city of Chicago as described above. The estimates for 4 of the 100 age groups and for the city as a whole are given in Table 4-3.

Soiling and Corrosion of Materials

Barrett and Waddell, among other authors, separate the effects of air quality improvements on inanimate materials into soiling and a residual category termed corrosion, or sometimes generalized material damage.[18] This corresponds to the physical science distinction between deposition of particles on surfaces and chemical interaction between materials and pollutants that causes decay of the material. Benefit evaluation is based on the observation that in either case the damage is to a durable material that is typically subjected to maintenance at periodic intervals. An improvement in air quality makes it possible to reduce the frequency of maintenance with no detriment to the level of service provided by the material.

The dollar benefit per unit of materials can be estimated by determining how much the maintenance frequency can be reduced, and then comparing present values of the streams of maintenance expenditures before and after the improvement. The aggregate benefit is computed by multiplying the benefits per unit by the total number of units of material subject to the air quality effects of the policy, and summing over the kinds of materials considered.

This analysis has been done for 5 common household maintenance operations. These include exterior and interior painting of residences, assumed to be related to the level of sulfur dioxide in the atmosphere, and cleaning of

Table 4-3. Health Care Benefits

Color-Sex Group	*Age in 1972 (Years)*	*Benefits Per Person (Dollars)*	*Total, City of Chicago (Millions of Dollars)*
White Males	Newborn	5.30	
	35	4.80	
	60	3.30	
	85	1.10	
	All ages		4.7
White Females	Newborn	5.30	
	35	5.00	
	60	3.70	
	85	1.20	
	All ages		5.2
Negro Males	Newborn	5.10	
	35	4.60	
	60	3.10	
	85	1.60	
	All ages		2.4
Negro Females	Newborn	5.20	
	35	4.70	
	60	3.50	
	85	1.70	
	All ages		3.3
Total			$15.6 Million
Per Household			$13.70

Note: Figures are 1972 present values of reductions in health care expenses, corresponding to a one $\mu g/m^3$ reduction in SO_2 or particulate levels. A 10 percent discount rate was used.

clothing, draperies, windows, and other surfaces, assumed to be affected mainly by the level of particulates. Estimates of average frequencies of maintenance in urban areas were obtained from existing literature on material damages of air pollution.[19] Changes in these frequencies were estimated by assuming that each frequency is proportional to the level of air pollution. The resulting aggregate benefit estimates per household appear in Table 4-4.

Expenditures Diverted from Other Uses

As discussed in the introduction to this chapter, it is likely that an individual would be willing to forego some of his current and future consumption in addition to giving up the gains discussed above in exchange for an air quality improvement. There is no information available for making an informed judgment as to the magnitude of this component of willingness to pay. The figure used in conjunction with the Chicago fuel policy analysis is based on the assumption that an extension of the expected duration of life by 1 percent would result in a willingness to forego 1 percent of the total present value of all future expenditures. Data in Tables 4-1 and 4-5 are used to make the computa-

Table 4-4. Household Maintenance Benefits (Dollars)

	Reduction in:	
Type of Maintenance	*Particulates*	*Sulfur Dioxide*
Exterior Painting of Residences	0.0	3.60
Interior Painting of Residences	0.0	12.40
Laundry, Frequent Cleaning of Interior Surfaces	30.60	0.0
Dry Cleaning and Other Infrequent Cleaning	5.30	0.0
Window Cleaning	9.50	0.0
Total per Household	45.40	16.00

Note: Figures are 1972 present values of reductions in household maintenance costs, corresponding to a one $\mu g/m^3$ reduction in pollution level as indicated. A 10 percent rate was used.

tion. For example, consider a newborn white male. Table 4-5 shows that 1 percent of the present value of his expected future expenditures is $453. (This is estimated in a present value computation involving the survival probabilities and expenditure assumptions discussed above, with a 2 percent real growth rate of yearly expenditure level, discounting future values back to the present at 10 percent.) Table 4-1 shows that his expected longevity is 65.7 years; Table 4-5 indicates that 1 percent of this is 240 days. The estimated willingness to divert expenditures from other uses is therefore calculated at the rate of $453 per 240 days of increase in expected longevity. Table 4-1 also shows that a one $\mu g/m^3$ improvement in suspended particulate levels extends expected longevity by 1.9 days. The corresponding willingness to divert income from other uses is $3.58 (453 × 1.9 / 240), as shown in Table 4-5.

Summary of Benefit Estimates

Table 4-6 consolidates the estimates of willingness to pay per household for a one $\mu g/m^3$ reduction in sulfur dioxide or particulates. The figures are present values; corresponding annualized amounts are a little more than 10 percent of the numbers shown. Some of the benefits for sulfur dioxide reductions could not be evaluated because of the lack of existing damage statistics. The table shows that the benefits in the categories related to current expenditures that continue through a person's life greatly outweigh benefits related to extension of the length of life. For particulates, about 5 percent of the total benefits are from reduced maintenance of materials, about 15 percent from reduced health care, 20 percent from diverted expenditures, and the rest is for increased longevity.

Table 4-5. Estimated Expenditures Diverted from Other Uses

		Basis of Evaluation		Present Value	
Color-Sex Group	*Age in 1972 (Years)*	*1% of Expected Remaining Longevity (Days)*	*1% of Present Value of Future Expenditures (Dollars)*	*Per Person (Dollars)*	*Total, City of Chicago (Millions of Dollars)*
White Males	Newborn	240	453	3.58	
	35	125	414	4.76	
	60	53	281	4.82	
	85	14	107	3.19	
	All Ages				4.61
White Females	Newborn	264	459	4.29	
	35	145	432	5.71	
	60	66	278	7.39	
	85	14	107	5.81	
	All Ages				6.83
Negro Males	Newborn	219	252	4.93	
	35	114	226	5.89	
	60	52	155	5.45	
	85	20	80	3.72	
	All Ages				2.74
Negro Females	Newborn	238	255	7.24	
	35	127	234	9.69	
	60	61	171	10.23	
	85	21	84	7.10	
	All Ages				5.67
Total					$19.85 Million
Per Household					$17.50

Note: The figures are 1972 present values corresponding to a one $\mu g/m^3$ reduction in particulate level.

Table 4-6. Summary of Benefits in Specific Damage Categories (Dollars)

Benefit Category	*Present Value per Household of a One $\mu g/m^3$ Reduction in:*		*Reference*
	Particulates	*Sulfur Dioxide*[a]	
Life Expectancy:			
Extension of Earnings	5.40	0.0	Table 4-2
Defer Expenses of Final Illness and Burial	.31	0.0	Table 4-2
Reduced Health Care Expenses	14.00	14.00	Table 4-3
Reduced Frequency of Materials Maintenance	45.00	16.00	Table 4-4
Expenditures Diverted from Other Uses	17.00	0.0	Table 4-5
Total	82.00	30.00	

[a]The zero entries for SO_2 reflect an absence of adequate statistical data on the rate at which SO_2 reduction reduces death rates. A 10 percent discount rate was used throughout the analysis.

This analysis has focused on selected air pollution damages in the residential sector to persons, households, and residential properties. A complete investigation would also quantify benefits in the following areas:

Improved visibility:
- Reduction in traffic accidents.
- Diminished use of electricity.

Reductions in damages to crops and ornamental vegetation.
Improvement in animal health.
Less damage to goods in retail stores.
Less damage to works of art.
Less corrosion damage to industrial equipment and other industrial and commercial properties.
Reduction in the adverse effects of pollution in the upper atmosphere.[20]

The dollar values of the first three of these are small; this is demonstrated adequately in the studies cited. However, the situation is less clear for the others; they have been neglected only because of an absence of adequate information which would permit making an estimate of the associated benefit of a unit improvement in air quality. A discussion of possible magnitudes of some of these other benefits is given in Appendix H.

BENEFIT EVALUATION USING FACTOR REWARDS APPROACH

Another method of estimating a family's willingness to pay for a given air quality improvement is to rely on analyses of spatial variations in urban property values, rather than adding up estimates related to itemized health and material damages. Numerous studies have demonstrated that there is an inverse relationship between air pollution levels and market values of single-family dwellings, which take into account the influence of other factors affecting these values. In effect, the costs of localized damages within an urban area are passed on to the property owners in the form of lower property values. By observing and analyzing the property values, one can estimate the values of the damages which they reflect.

The results of Crocker's study of sale prices of more than 1000 FHA-insured dwellings in Chicago are especially of interest here.[21] His analysis implies that a difference of one $\mu g/m^3$ of suspended particulates affects residential property values by at least \$1.75, and possibly as much as \$3, per \$1000 of property value. Since property value is essentially a market estimate of the expected value of future housing services, these results are interpreted as revealing market estimates of the present value of benefits from a reduction in particulates of one $\mu g/m^3$. In some of his statistical experiments, no significant additional effect of spatial variation in sulfur dioxide levels was revealed, but in others, it is suggested that the effect of a one $\mu g/m^3$ difference in sulfur dioxide is around \$4 per \$1000 of residential property value.

Two sets of estimates will be utilized in the discussion of benefits of the Chicago fuel policies. The first is \$1.75 per $\mu g/m^3$ for particulates and no effect for sulfur dioxide. This corresponds to the smallest of Crocker's results. The second is \$2.50 for particulates and \$1.75 for SO_2. The \$2.50 is a representative estimate for particulates, while the \$1.75 is a conservative estimate, among the sulfur dioxide results that did indicate some magnitude of damage. Taking the average value of a housing unit in Chicago of \$20,500,[22] the social benefits of a one $\mu g/m^3$ air quality improvement for a household can be calculated. The figures are shown in Table 4-7, which also includes corresponding figures based on the direct analysis of the impacts on health and materials.

As can be seen in the table, the direct method gives estimates that are greater than those implied in the property value studies. One important reason for this is that urban property value differentials probably reflect only the portion of pollution damages which are readily apparent to people choosing where to live. On the other hand, the direct analysis itself uses a number of very rough coefficients and may have produced an overestimate. Therefore, in the analysis to follow, three estimates of fuel policy benefits are presented, based on the three sets of coefficients in Table 4-7. The favored of these is the middle estimate; it corresponds to the best estimates from the property

Table 4-7. Benefit Estimates Using Two Alternative Methods (Dollars)

	Reduction in Particulates		*Reduction in Sulfur Dioxide*	
Basis of Estimation	*Per $1000 of Residential Property*	*Per Household*	*Per $1000 of Residential Property*	*Per Household*
Property Value Method				
Minimum Estimate	1.75	36	0.0	0
Mid-range Estimate	2.50	51	1.75	36
Specific Damages Method (from Table 4-6)	4.00	82	1.45	30

Note: All figures are present values for an immediate one $\mu g/m^3$ improvement in air quality, continuing indefinitely into the future.

value studies, and it is about half the estimate yielded in the direct analysis. (Appendix H presents an evaluation of this approach, which considers the possible impact of non-residential emission controls on marginal benefits and compares the benefit values of this study with those developed by other authors.)

CHICAGO FUEL POLICY BENEFITS

Two methods have been discussed for estimating the social benefit of an air quality improvement to households in an urban area. One utilizes data on the total number of households affected by the policy, together with the estimated benefits per household from Table 4-7. The other uses data on the value of residential property affected by the policy in conjunction with the figures on benefit per $1000 of property value in the table. To investigate the effects of residential fuel policies in Chicago, the second approach has been used.

The benefits were computed separately for each square mile cell in the city of Chicago and summed to produce a city-wide benefit estimate. Aggregate residential property values for each cell were estimated using the 1970 U.S. Census of Housing data for census tracts.[23] The aggregate value of owner-occupied and vacant for-sale dwellings in each tract is reported by the Census Bureau. The aggregate monthly contract rent of renter-occupied and vacant for-rent dwellings is also reported. The latter can be converted into an equivalent total value, assuming a 10 percent discount rate, by multiplying it by 12, the number of months per year, and then by 11, a capitalization factor. (The present value of a succession of yearly $1 payments extending indefinitely into the future is $1 + 1/(1+r) + 1/(1+r)^2 + 1/(1+r)^3 + \ldots = (1+r)/r = 11$ when $r = 0.10$.) In both cases, the census figures relate only to the housing units for which values or rents were reported by occupants or owners in their census questionnaires. If the average values of units for which values were reported in the census are the same as the average values of the other units, the aggregate benefit estimates would be about 15 percent higher than the figures actually calculated. (In the *1970 U.S. Census of Housing*, the total number of dwelling units was about 15 percent higher than the sum of the number of dwelling units for which value was tabulated plus the number for which rent was tabulated, in the Chicago area.[24])

Tables 4-8 and 4-9 contain aggregate benefit estimates for the fuel policies, for each of the three alternative rates of benefit valuation provided in Table 4-7, for 10 percent and 20 percent discount rates, respectively. Considering the mid-range estimates for the city of Chicago, the benefits of the Chicago low sulfur ordinance are approximately $20 million annually. Banning coal by 1975 raises the benefits to about $25 million annually. Banning both coal and oil for residential space heating raises the benefits to about $28.5 million annually. The minimum and maximum estimates indicate that the benefits of the low sulfur law range between $1 and $20 million annually. A

Table 4-8. Benefits for Five Policies, By Location, Using a 10 Percent Discount Rate (Millions of Dollars)

Policy Comparison	*Chicago*		*Rest of 8-County Area*		*Total*	
	P.V.	*Ann.*	*P.V.*	*Ann.*	*P.V.*	*Ann.*
Minimum Valuation						
Sulfur Law vs. No Control	9.7	1.0	2.1	0.21	12.	1.2
Coal Ban 1975 vs. Sulfur Law	17.	1.9	4.4	0.49	21.	2.4
Coal & Oil Ban 1975 vs. Coal Ban 1975	7.1	0.79	2.1	0.23	9.2	1.0
Coal Ban 1977 vs. Coal Ban 1975	(5.7)	(0.63)	(1.4)	(0.15)	(7.1)	(0.78)
Coal & Oil Ban 1977 vs. Coal & Oil Ban 1975	(7.1)	(0.79)	(1.9)	(0.21)	(9.0)	(1.0)
Mid-Range Valuation						
Sulfur Law vs. No Control	194.	20.	41.	4.3	240.	24.
Coal Ban 1975 vs. Sulfur Law	51.	5.6	13.	1.5	64.	7.1
Coal & Oil Ban 1975 vs. Coal Ban 1975	27.	2.9	7.9	.87	34.	3.8
Coal Ban 1977 vs. Coal Ban 1975	(17.)	(1.9)	(4.1)	(.45)	(21.)	(2.4)
Coal & Oil Ban 1977 vs. Coal & Oil Ban 1975	(22.)	(2.5)	(6.0)	(.67)	(28.)	(3.1)
Maximum Valuation						
Sulfur Law vs. No Control	170.	18.	36.	3.8	210.	22.
Coal Ban 1975 vs. Sulfur Law	61.	6.7	16.	1.8	77.	8.5
Coal & Oil Ban 1975 vs. Coal Ban 1975	30.	3.3	8.9	.99	39.	4.3
Coal Ban 1977 vs. Coal Ban 1975	(21.)	(2.3)	(4.9)	(.54)	(25.)	(2.8)
Coal & Oil Ban 1977 vs. Coal & Oil Ban 1975	(26.)	(2.9)	(7.2)	(.80)	(33.)	(3.7)

Note: P.V. = present value; Ann. = annualized from 1973 (1969 for sulfur law vs. trends) to 1990. Figures in parentheses represent negative benefits.

similarly wide range of benefit estimates is also indicated for the other policies.

The minimum valuation assigns no benefits to sulfur dioxide reductions, the main impact of the sulfur law. It is of interest that the value of particulate reduction associated with the sulfur law is about $1 million annually. This benefit is associated with conversions from coal motivated by the sulfur law.

Table 4-9. Benefits for Five Policies, By Location, Using a 20 Percent Discount Rate (Millions of Dollars)

Policy Comparison	*Chicago*		*Rest of 8-County Area*		*Total*	
	P.V.	*Ann.*	*P.V.*	*Ann.*	*P.V.*	*Ann.*
Minimum Valuation						
Sulfur Law vs. No Control	7.6	1.3	1.6	.27	9.2	1.6
Coal Ban 1975 vs. Sulfur Law	21.	3.6	5.4	.93	26.	4.5
Coal & Oil Ban 1975 vs. Coal Ban 1975	8.1	1.4	2.4	.42	10.	1.8
Coal Ban 1977 vs. Coal Ban 1975	(8.4)	(1.5)	(2.0)	(.35)	(10.)	(1.8)
Coal & Oil Ban 1977 vs. Coal & Oil Ban 1975	(10.)	(1.8)	(2.8)	(.49)	(13.)	(2.3)
Mid-Range Valuation						
Sulfur Law vs. No Control	210.	35.	44.	7.5	250.	43.
Coal Ban 1975 vs. Sulfur Law	63.	11.	16.	2.8	79.	14.
Coal & Oil Ban 1975 vs. Coal Ban 1975	30.	5.2	8.9	1.5	39.	6.7
Coal Ban 1977 vs. Coal Ban 1975	(25.)	(4.4)	(6.0)	(1.0)	(31.)	(5.4)
Coal & Oil Ban 1977 vs. Coal & Oil Ban 1975	(33.)	(5.6)	(8.9)	(1.5)	(41.)	(7.2)
Maximum Valuation						
Sulfur Law vs. No Control	190.	32.	39.	6.7	230.	38.
Coal Ban 1975 vs. Sulfur Law	75.	13.	20.	3.3	94.	16.
Coal & Oil Ban 1975 vs. Coal Ban 1975	34.	5.9	10.	1.8	44.	7.6
Coal Ban 1977 vs. Coal Ban 1975	(30.)	(5.2)	(7.2)	(1.2)	(37.)	(6.5)
Coal & Oil Ban 1977 vs. Coal & Oil Ban 1975	(39.)	(6.7)	(10.)	(1.8)	(49.)	(8.5)

Note: P.V. = present value; Ann. = annualized from 1973 (1969 for sulfur law vs. trends) to 1990. Figures in parentheses represent negative benefits.

Tables 4-8 and 4-9 also present numbers that make it possible to assess the reduction in air quality improvement benefits from delaying the coal ban or the coal and oil ban from 1975 until 1977. Compared with a benefit of $5.6 million annually in Chicago for the 1975 coal ban, a 1977 coal ban would provide a benefit of only $3.7 million. That is, a reduction in benefits (a cost) of $1.9 million results from delaying the coal ban for two years. There is a similar reduction in benefits from delaying a coal and oil ban from 1975 to 1977, amounting to $2.5 million annually.

Table 4-10. Benefits per Household for Five Policies, By Location, Using a 10 Percent Discount Rate (Dollars)

Policy Comparison	*Chicago Average*		*Rest of 8-County Area, Average*		*Uptown, Chicago*	
	P.V.	*Ann.*	*P.V.*	*Ann.*	*P.V.*	*Ann.*
Minimum Valuation						
Sulfur Law vs. No Control	9.20	.96	2.00	.21	15.00	1.60
Coal Ban 1975 vs. Sulfur Law	16.00	1.80	4.31	.48	28.00	3.10
Coal & Oil Ban 1975 vs. Coal Ban 1975	6.70	.75	2.10	.23	8.50	.95
Coal Ban 1977 vs. Coal Ban 1975	(5.40)	(.60)	(1.30)	(.14)	(9.20)	(1.00)
Coal & Oil Ban 1977 vs. Coal & Oil Ban 1975	(6.71)	(.74)	(1.80)	(.20)	(11.00)	(1.20)
Mid-Range Valuation						
Sulfur Law vs. No Control	190.00	19.00	40.00	4.20	350.00	36.00
Coal Ban 1975 vs. Sulfur Law	48.00	5.40	13.00	1.50	90.00	10.00
Coal & Oil Ban 1975 vs. Coal Ban 1975	25.00	2.80	7.70	.85	36.00	4.00
Coal Ban 1977 vs. Coal Ban 1975	(16.00)	(1.80)	(3.80)	(.42)	(29.00)	(3.20)
Coal & Oil Ban 1977 vs. Coal & Oil Ban 1975	(21.00)	(2.30)	(5.70)	(.63)	(36.00)	(4.00)
Maximum Valuation						
Sulfur Law vs. No Control	160.00	17.00	35.00	3.70	310.00	32.00
Coal Ban 1975 vs. Sulfur Law	58.00	5.50	16.00	1.70	110.00	12.00
Coal & Oil Ban 1975 vs. Coal Ban 1975	29.00	3.20	8.70	.97	39.00	4.30
Coal Ban 1977 vs. Coal Ban 1975	(19.00)	(2.20)	(4.80)	(.53)	(34.00)	(3.80)
Coal & Oil Ban 1977 vs. Coal & Oil Ban 1975	(25.00)	(2.80)	(7.00)	(.78)	(42.00)	(4.60)

Note: P.V. = present value; Ann. = annualized from 1973 (1969 for sulfur law vs. trends). Figures in parentheses represent negative benefits. Source: Derived from Table 4-8 and from geographically detailed data underlying Table 4-8.

The tables also indicate something about the distribution of benefits over the Chicago area. First, there is an air quality improvement and corresponding benefit outside the boundaries of the city as a result of the policy, even though coal burning is largely confined to within Chicago itself. The additional benefits in the suburban area are approximately 20 percent of the benefits within the city, for most of the policy options. Table 4-10 presents figures on benefits per household. The last two columns contain estimates for the Uptown area of Chicago's north side, where the air quality impact of the fuel policies is the highest. The benefit there is about 1.5 to 2 times the average benefit per household in the city as a whole. The table also indicates that the average benefit to a household in the suburban area is about one quarter the average benefit per household in the city.

Additional benefit calculations were made for the 1975 coal ban assuming that premature abandonments or an exogenous gas price increase occur (see Chapter 2 for details). Abandonments affect benefits by altering the linear conversion path assumed for the coal ban. This has the effect of reducing the impact of the coal ban in 1973 and 1974. An exogenous gas price increase would have the effect of reducing the rate of conversion assumed under the low sulfur law. Therefore, it would increase the impacts of the coal ban. Table 4-11 shows the results of these analyses.

The mid-range estimates from Tables 4-8 and 4-9 are recapitulated in Table 4-12. These values are the basic benefit estimates to be compared with Chapter 2's cost estimates in the following chapters.

NOTES TO CHAPTER FOUR

1. This approach has been used in the context of environmental–economic analysis by Ridker, Crocker, and others. See R.B. Ridker, *Economic Costs of Air Pollution: Studies in Measurement* (New York: Praeger Publishers, 1967), p. 18; and T.D. Crocker, *Urban Air Pollution Damage Functions: Theory and Measurement*, National Technical Information Service Publication No. PB–197668 (Springfield, Va., 1970), p. 23.
2. T.C. Schelling, "The Life You Save May Be Your Own," in *Problems in Public Expenditure Analysis*, ed. S.B. Chase (Washington: The Brookings Institution, 1968).
3. U.S. Office of the Federal Register, *Code of Federal Regulations*, Vol. 40, *Protection of Environment* (Washington: U.S. Government Printing Office, 1972), p. 60.
4. This taxonomy of modes of evaluation is from P.W. Taylor; see his *Normative Discourse* (Englewood Cliffs, N.J.: Prentice-Hall, Inc., 1961), pp. 5–9, 32–33.
5. L.B. Lave and E.B. Seskin, "Does Air Pollution Shorten Lives?" in *Proceedings of the Second Research Conference of the Inter-University Committee on Urban Economics*, pp. 293–328.
6. The Lave-Seskin results confirm numerous clinical studies. For a summary, see U.S. National Air Pollution Control Administration, *Air Quality Criteria for Particulate Matter*, Publication No. AP–49 (Washington: U.S. Government Printing Office, 1969).
7. Since the Lave-Seskin study does not report results for sulfur dioxide, it was necessary to restrict this part of the work to particulates.

Table 4-11. Parametric Analysis of the Benefits of the Coal Ban Relative to the Sulfur Law, for the Chicago Area (Millions of Dollars)

				10% Discount Rate		*20% Discount Rate*	
	Year of Coal Ban	*Premature Abandonments*	*Exogenous Price Increase Adjustment Factor*	*Present Value*	*Annualized Benefits*	*Present Value*	*Annualized Benefits*
Minimum Valuation	1975	No	1.000	21	2.4	26	4.5
	1975	Yes	1.000	18	2.0	22	3.8
	1975	No	.768	29	3.3	35	6.1
	1975	No	.406	41	4.5	47	8.2
	1977	No	1.000	14	1.6	16	2.7
	1977	Yes	1.000	12	1.3	12	2.1
Mid-Range Valuation	1975	No	1.000	64	7.1	79	14.0
	1975	Yes	1.000	56	6.2	66	11.0
	1975	No	.768	89	9.9	110	18.0
	1975	No	.406	120	14.0	140	25.0
	1977	No	1.000	43	4.7	48	8.6
	1977	Yes	1.000	35	3.9	37	6.4
Maximum Valuation	1975	No	1.000	77	7.5	94	16.0
	1975	Yes	1.000	67	7.4	79	14.0
	1975	No	.768	110	12.0	130	22.0
	1975	No	.406	150	16.0	170	30.0
	1977	No	1.000	52	5.7	57	9.5
	1977	Yes	1.000	42	4.7	44	7.7

Table 4-12. Summary: Benefits of Fuel Policies Under the Most Likely Assumptions (Millions of Dollars)

		10% Discount Rate		*20% Discount Rate*	
	Estimation Period	*Present Value*	*Annualized Benefits*	*Present Value*	*Annualized Benefits*
Sulfur Law vs. No Control	1969–1990	240	24.0	250	43.0
Coal Ban 1975 vs. Sulfur Law	1973–1990	64	7.1	79	14.0
Coal & Oil Ban 1975 vs. Coal Ban 1975	1973–1990	34	3.8	39	6.7
Coal Ban 1977 vs. Sulfur Law	1973–1990	43	4.7	48	8.6
Coal & Oil Ban 1977 vs. Coal Ban 1977	1973–1990	27	3.1	29	4.9

Note: The figures are the mid-range estimates from Tables 4–8 and 4–9. Present values are for 1973.

8. Ridker, *op. cit.*, pp. 35–36.
9. U.S. Bureau of the Census, *1970 Census of Population*, Final Report PC(1)–D15, *Detailed Characteristics, Illinois* (Washington: U.S. Government Printing Office, 1972), pp. 1264–1265.
10. *Ibid.*, p. 1283.
11. The 0.9 is the average propensity to consume as indicated in studies of the relationship between aggregate personal consumption and personal disposable income from year to year in the U.S. See, for example, E. Shapiro, *Macroeconomic Analysis* (New York: Harcourt, Brace, and World, Inc., 1966), pp. 196–203.
12. The detailed formulas used to make all estimates presented in this section are discussed in Appendix H.
13. J. Mitford, *The American Way of Death* (New York: Simon and Schuster, 1963), pp. 39–41.
14. R.A. Schrimper, "Investigation of Morbidity Effects," Mimeographed (Chicago: University of Chicago, 1973), pp. 13–14.
15. L.B. Lave and E.P. Seskin, "Air Pollution and Human Health," *Science*, Vol. 139, No. 3947 (1970), p. 730.
16. Schrimper, *op. cit.*, p. 20. The estimate includes value of time lost, based on $10 per day per person.
17. Lave and Seskin, "Air Pollution and Human Health," *op. cit.*, p. 730.
18. L.B. Barrett and T.E. Waddell, *Cost of Air Pollution Damage: A Status Report*, U.S. Environmental Protection Agency Publication No. AP–85 (Research Triangle Park, N.C., 1973).
19. R.L. Salmon, *Systems Analysis of the Effects of Air Pollution on Materials*, National Technical Information Service Publication No. PB–209192 (Springfield, Va., 1972).
20. Lists of types of damage and estimates of their magnitudes can be found in many places. See, for example, J.J. O'Connor, Jr., *The Economic Cost of the Smoke Nuisance to Pittsburgh*, Mellon Institute of Industrial Research Smoke Investigation Bulletin No. 4 (Pittsburgh: University of Pittsburgh, 1913); Great Britain, *Parliamentary Papers*, Vol. 8 (1953–54) (*Reports from Commissioners, Inspectors, and Others*, vol. 1), "Report of the Committee on Air Pollution," Cmd. 9322, 1954; and Barrett and Waddell, *op. cit.*
21. Crocker, *op. cit.*
22. Estimated from Census Bureau data on aggregate rents, property values, and the numbers of rental and owner units in the city; see U.S. Bureau of the Census, *1970 Census of Housing*, Final Report HC(2)–44, *Metropolitan Housing Characteristics, Chicago, Ill. SMSA* (Washington: U.S. Government Printing Office, 1972), pp. 53, 54.
23. U.S. Bureau of the Census, *1970 Census of Population and Housing*, Fourth Count Summary Computer Tapes for Illinois.
24. *Ibid.*

Chapter Five

Distribution of Benefits and Costs

INTRODUCTION

Two types of analysis have been identified as necessary for determining the best restriction on fuels for residential heating. First, total benefits must be compared with total costs. Figures required to make such a comparison for a variety of fuel policies have been presented in Chapters 2 through 4. Second, any policy identified as optimal on the basis of total cost and benefit comparisons should be further examined for the distribution of its impacts among geographic areas and sectors of the economy. Inequities in the distribution of the economic impacts must be considered to ensure that specific groups of people are not bearing excessive costs for the benefit of others. If this occurs, a program to alleviate the inequities (e.g., subsides and low interest loans) may be called for. If the administrative costs of these programs do not change the sign of the net benefits of the policy, then the policy and the supporting income redistribution program is still justifiable.

In this study, the 1975 coal ban is identified as the best policy on the basis of comparing total benefits and costs. This conclusion is discussed in detail in the next chapter. In this chapter the distribution of the costs and benefits from the 1975 coal ban is considered from two different viewpoints. The first is the geographical distribution of benefits and costs over the city. This distribution is useful in identifying incidence of benefits, costs, and net benefits among income, race, and other socio-economic groups. The second is the functional distribution of costs and benefits–i.e., the distribution among economic sectors. The sectors that bear costs are fuel distributors and their employees; landlords, tenants, and homeowners; janitors; and non-residential property owners. Benefits accrue to the household sector–i.e., to individuals who reside in the region. Additional benefits accrue to property owners because of reduced material damages.

In the discussion that follows, the distribution of the economic impacts of the 1975 coal ban is analyzed employing the following set of assumptions: (1) the mid-range benefits, (2) no exogenous price increases, (3) an endogenous price increase determined by a winter gas supply elasticity of 3, (4) conversion costs twice those presented in Table 2-2, and (5) massive premature abandonments. It should be emphasized that this is not the most likely set of assumptions. The difference between the most likely assumptions, given at the end of Chapter 2, and the assumptions for this distributional analysis are that conversion costs are doubled and that the policy causes premature abandonments. The assumptions in the present chapter were chosen as possible reasons for adverse distributional effects, in order to provide a strong test of whether these effects could be important.

GEOGRAPHICAL DISTRIBUTION

The total social benefits of the 1975 coal ban were estimated by summing figures on a square mile basis throughout the city. Social costs for conversions, fuel price changes, abandonments, and displacements were calculated for the city as a whole. They were apportioned to each square mile based on the percentage of coal- or gas-using dwelling units in each grid in order to determine where the net benefits are positive or negative, and whether the policy favors any socio-economic group. The ratios of benefits to costs resulting from this analysis are indicated in Figure 5-1, for each square mile where annual costs or annual benefits are greater than $5,000. The figure shows that in most sections of the city with appreciable levels of benefits or costs, positive net benefits can be expected from the policy. Costs exceed benefits in only 16 of the 172 cells for which estimates are given. In 54 of the areas, the benefit-cost ratio is greater than 2, and in many instances, it is substantially higher.

Figure 5-2 shows income levels by geographical area within the city. Three of the 16 areas where benefits are less than or equal to costs have income levels substantially below average. These are located near Lake Michigan in the central portion of the city. One of the 16, on the northern border of the city, is in a relatively high-income neighborhood. The others are in neighborhoods where the income level is average or slightly below average. This latter group includes Uptown, where the impact of the coal ban on air quality levels is the greatest. In this area benefits are about equal to costs. Of the ten square-mile areas in Figure 5-2 with average income levels less than $6,000, benefits are less than costs in the three lakeshore and central locations mentioned previously. Benefits are substantially greater than costs in five of these low income areas in the south and central parts of the city. Benefits are slightly greater than costs in one other low income area.

There are 36 square-mile areas of the city in which average income levels are greater than $12,000 per year. Of these, one quarter are not expected

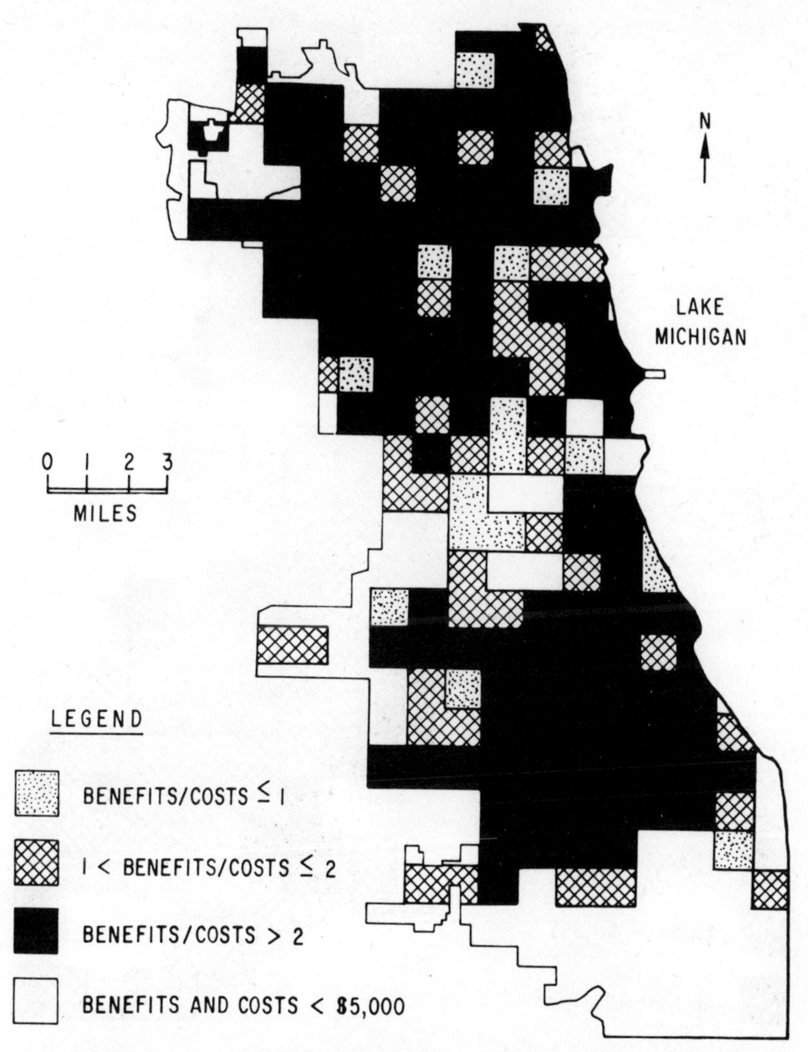

Figure 5-1. Benefit-Cost Ratios by Square Mile in Chicago.

to receive appreciable benefits or bear appreciable costs as a result of the coal ban. One square mile, as previously mentioned, has positive net costs. The others have benefits greater than costs. Positive net benefits are also expected in many areas where income levels are average or slightly below average.

Figure 5-3 indicates the racial composition of neighborhoods of the city. Thirty-five square-mile areas had at least 80 percent non-white population in 1970. Of these, two (the central lake-shore areas previously identified as having relatively low-income levels) have positive net costs, and two do not

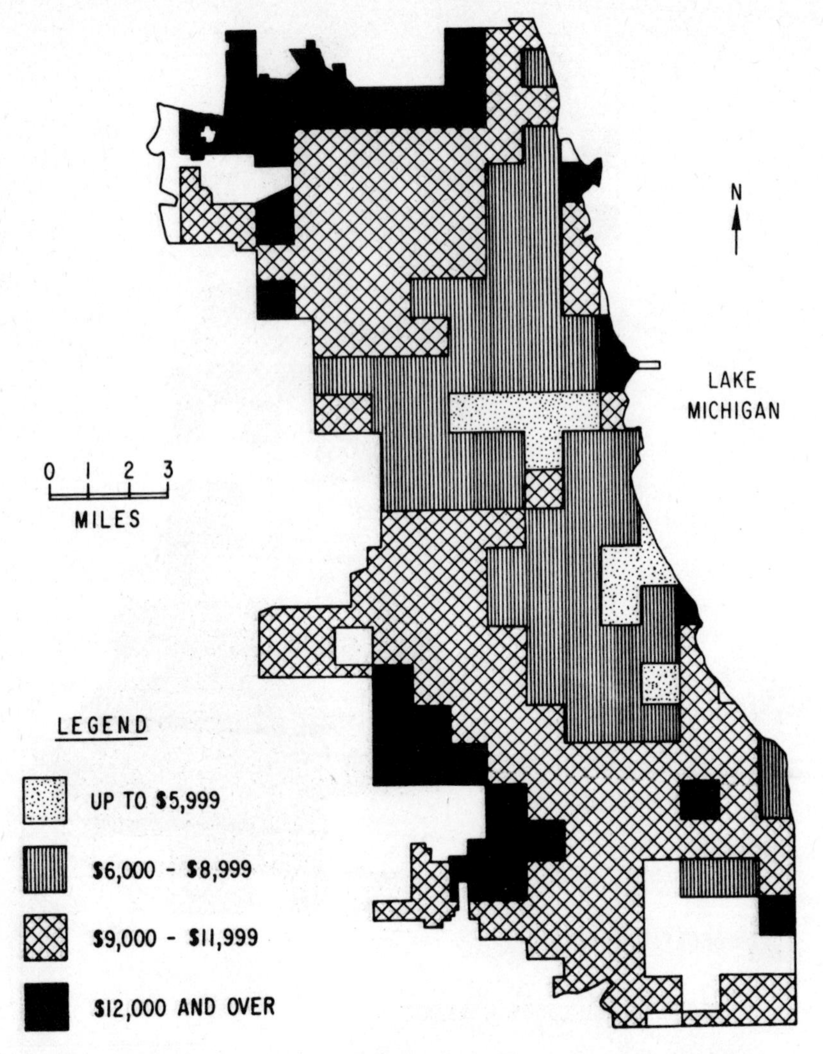

Figure 5–2. Income Levels by Square Mile, Chicago, 1970.

have substantial magnitudes of either benefits or costs. The others, comprising almost 90 percent of the total, are expected to receive positive net benefits as a result of the coal ban. In addition, positive net benefits are expected in many, although not all, of the largely white neighborhoods.

Based on these observations, it can be concluded that the policy does not systematically favor either high-income or low-income areas. The policy also is substantially without any racial bias. With few exceptions, predominantly non-white areas of the city have positive net benefits.

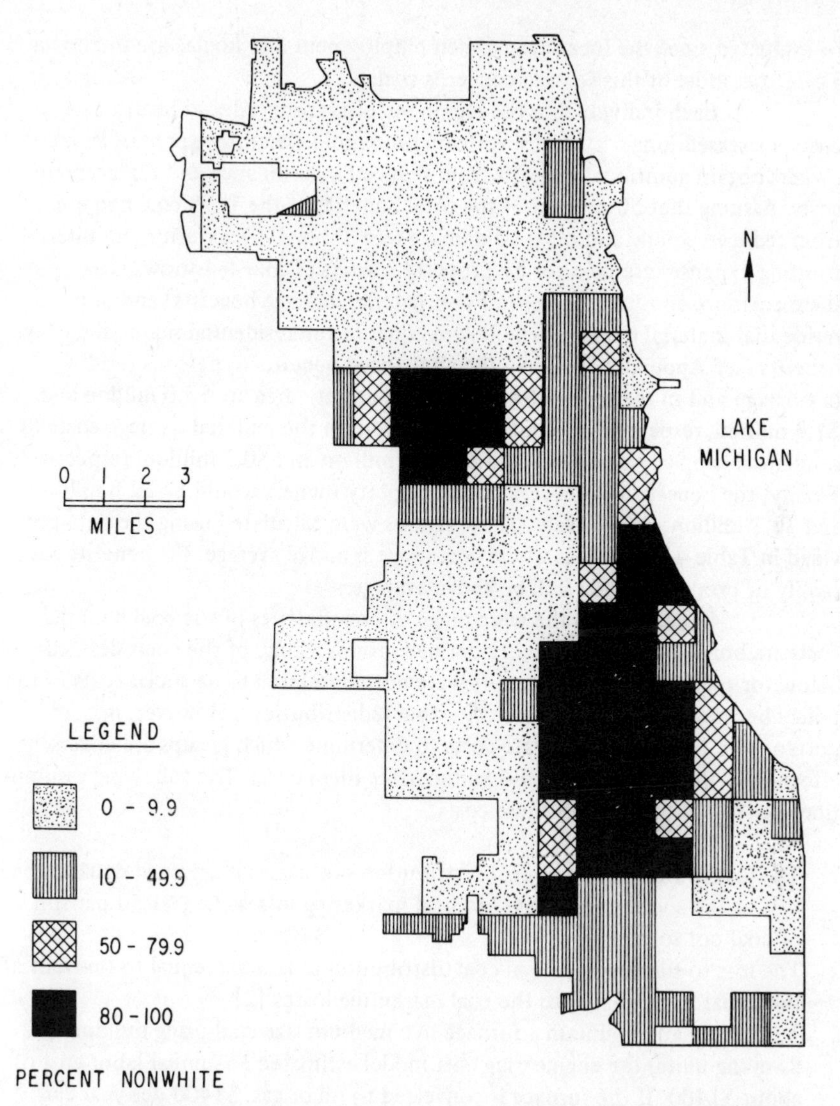

Figure 5-3. Racial Composition by Square Mile, Chicago, 1970.

FUNCTIONAL DISTRIBUTION

In this section the benefits and costs to individuals in various economic sectors are evaluated for the 1975 coal ban. The direct benefits they obtain depend on the air quality improvements at their residential locations and places of work. Except for some general comments, the benefits to each economic sector can not be evaluated since the locations of their employment and homes are unknown. Therefore, most of this section concerns costs.

Each individual in the region benefits from reduced health care costs, increased longevity and savings on household cleaning expenses. Property owners obtain additional benefits from reduced maintenance and replacement costs. Assume that 50 percent of the total benefits of the 1975 coal ban are from reduced soiling and material damage (see Table 4–6), exterior and interior painting expenses are paid for by property owners (Table 4–4 shows that these costs are about 23 percent of the material damage benefits) and non-residential material damage benefits are equal to the residential material damage benefits (see Appendix H). Then the annualized benefits to persons residing in Chicago and in the remainder of the eight-county area are $5.0 million and $1.3 million, respectively. Similarly, reductions in the material damage costs to residential property owners would be $0.6 million and $0.2 million, respectively. Finally, the benefits to non-residential property owners would be $2.8 million and $0.7 million, respectively. These figures were calculated using the data provided in Table 4–8. In areas where coal usage is above average, the benefits per family or property owner will be above average.

In contrast to the wide range of beneficiaries of the coal ban, the costs are borne by more explicitly defined groups. Some of the costs defined below for specific economic sectors were not considered to be social costs of the policy because they only represent income redistributions. However, it is important to estimate these private costs to determine which groups are adversely affected by the policy. Table 5–1 summarizes these costs. The following assumptions were used to quantify these costs:

1. The loss to the owners of coal distribution businesses is assumed equal to one half of the social costs related to coal marketing losses—i.e., $1.50 per ton of coal not sold.[1]
2. The loss to the employees in coal distribution is assumed equal to one half of the social costs related to the coal marketing losses.[2]
3. To operate and maintain a furnace in a medium size coal-using building (22 dwelling units) the engineering cost model estimated an annual labor cost of about $1460. If the furnace is converted to oil or gas, $1400 per year can be saved in labor costs. This would be the private cost to the janitors if they had no alternative employment. Assuming that half of the janitors will not find alternative employment—i.e., they will be underemployed—these costs to janitors are about $7 per ton of coal or 4.6 times the private cost per ton of coal to coal distributors.
4. The losses to landlords and homeowners include the costs of conversions, rents lost due to abandonment, fuel savings (adjusting for endogenous price increase), and savings in janitorial labor. These costs can be calculated from the data in Tables F–9, 2–9, and 2–6. The savings in janitorial labor are twice

Table 5-1. Costs of the 1975 Coal Ban, By Economic Sector (Millions of Dollars)

	10% Discount Rate	*20% Discount Rate*
Coal Distributors	.29	.32
Employees in Coal Distribution	.29	.32
Janitors	1.33	1.45
Landlords and Homeowners	(.85)	(1.17)
Tenants Prematurely Displaced	.30	.44
Non-residential Winter Gas Consumers	.65	.72

Note: Costs are marginal with respect to the sulfur law. Figures in parentheses represent negative marginal costs–i.e., savings.

the loss to janitors because it was assumed that half of the janitors will find alternative employment.

5. The losses to the tenants prematurely displaced because of abandonments consist of moving and psychic costs. These annualized costs to tenants can be calculated from the data reported in Table 2-6.[3]
6. The loss to non-residential winter gas consumers is about equal to the endogenous price increase costs to residential gas consumers which are reported in Table 2-7.

The private costs presented in Table 5-1 have to be divided by the number of individuals in each economic sector to determine the burden placed on each individual. The number of tenants displaced and the number of non-residential winter gas users are large so that the per individual cost is less than $10 per year. Landlords and homeowners as a group obtain net benefits of about $150 per year per building. However, landlords who abandon their buildings would bear a net cost. The three groups that incur significant costs are janitors, owners of coal distribution businesses and employees in coal distribution. The cost per janitor per building is about $200 per year while coal distributors lose about $5,000 per year per distributorship. Employees in coal distribution (assuming 10 per coal yard) are losing about $500 per year.[4]

The distribution of the private costs associated with the 1975 coal ban is, in general, not regressive in the sense that costs are not borne disproportionately by the low-income sectors. Based on this analysis, however, an income compensation policy may be decided to be appropriate. As noted at the beginning of this chapter, assumptions for the distributional analysis were chosen with a view to identifying possible adverse effects rather than most likely effects. Under the most likely assumptions, there would be no abandonments with the result that tenants would bear no costs. Thus, the most likely outcome is that the adverse distributional impacts would be even less than estimated in this chapter.

NOTES TO CHAPTER FIVE

1. The mark-up of coal is about \$6.73/ton. It has been assumed that \$3 are returns to coal distributors, \$3 to employees, and \$.73 are variable costs that can be saved. It is assumed that \$1.50 of the owners' earnings per ton of coal are returns to entrepreneural skills and capital which are not adaptable to alternative uses.
2. As a parallel assumption to that in the previous footnote, half of the labor skills of employees in coal distribution are assumed to be non-transferable. Thus, the value of the lost income is assumed to be \$1.50 per ton of coal not sold.
3. If the tenants have to pay higher rents in their new locations, then these private costs must be added. In addition, these higher rents are private benefits to landlords.
4. Potential losses to coal miners were not evaluated. Since the demand for coal for space heating is a small percentage of the total demand for coal, the costs to coal miners will be very small.

Chapter Six

Summary and Conclusions

INTRODUCTION

One purpose of this chapter is to compare the total social cost and benefit estimates presented in Chapters 2 and 4 in order to reach an overall judgment regarding which policy will be most beneficial. In addition, the chapter considers broad issues of national concern raised by the study. These include (1) the appropriateness of the federal approach to environmental control and (2) the energy policy implications of a restriction on coal or oil usage in view of possible shortages of clean fuels.

TOTAL BENEFITS AND COSTS

The marginal social cost and benefit estimates using the basic assumptions are given in Table 6-1, which summarizes results from previous chapters. From these marginal comparisons, total net annual benefits can be calculated for two mutually exclusive policy scenarios, as presented in Table 6-2. The difference between the two scenarios is in the fuel ban compliance date, 1975 or 1977.

Consider the sequence of air quality improvements resulting from a sulfur law, the 1975 coal ban, and the 1975 coal and oil ban. Although each of the control policies provides a positive net benefit to society, the coal and oil ban maximizes net benefits. This is true for discount rates of 10 and 20 percent. Now consider the sequence of air quality improvements resulting from a sulfur law, the 1977 coal ban, and the 1977 coal and oil ban. Again, the coal and oil ban is the preferred policy. Furthermore, a compliance deadline of 1975 is preferred over a deadline of 1977 since the net benefits in 1975 are greater than those in 1977 for each fuel ban policy. This holds for the other parametric assumptions considered. Therefore, only the 1975 fuel ban policies will be considered in the remainder of this chapter.

Table 6-1. Marginal Costs and Benefits of Fuel Policies (Millions of Dollars)

Policy Comparison	*Annualized Costs (Savings)*		*Annualized Benefits*	
	(10%)	*(20%)*	*(10%)*	*(20%)*
Sulfur Law vs. No Control	0.6	0.4	24.0	43.0
Coal Ban 1975 vs. Sulfur Law	(0.7)	(.3)	7.1	14.0
Coal and Oil Ban 1975 vs. Coal Ban 1975	3.8	5.7	3.8	6.7
Coal Ban 1977 vs. Sulfur Law	(0.6)	(0.2)	4.7	8.6
Coal and Oil Ban 1977 vs. Coal Ban 1977	3.0	4.7	3.1	4.9

Note: These data are from Tables 2-10 and 4-12.

The initial comparison between costs and benefits suggests that the coal and oil ban is the best residential fuel policy. The marginal benefits and costs of banning oil in addition to coal are about equal under the basic assumptions. Therefore, the result that the coal and oil ban is optimal is highly sensitive to parametric changes. For example, if conversion costs are higher, abandonments occur, or the minimum estimate is the appropriate measure of benefits, then banning oil in addition to coal would not be justified. It is difficult to determine whether an increase or decrease in the price of gas relative to oil will make the policy more or less attractive, because both costs and benefits are affected. The result would be highly sensitive to the assumptions concerning conversions to or from oil and the use of oil in new buildings. On the other hand, if oil prices change relative to gas prices, or the maximum benefit estimate is appropriate, then the coal and oil ban would be more attractive. Because the sign of the estimated net marginal benefit of the coal and oil ban varies for reasonable parametric changes, banning oil in addition to coal can not be definitely recommended.

Benefits may exceed costs for an oil ban policy restricted to relatively large residential and non-residential sources using residual oils for space heating. Chapter 2 showed that a major component in the social cost of an oil ban to the residential sector is the cost of furnace conversion. It was shown that the larger the building, the lower the conversion costs per dwelling unit. Although fuel savings are also less, the relative reduction in costs is greater than the relative reduction in savings as building size increases. In addition, benefits of restricting oil use by large buildings are likely to be higher, per dwelling unit, than for smaller buildings, since in the former, highly polluting residual oil is typically used.

For the coal ban, marginal benefits exceed marginal costs by a

Table 6-2. Total Benefits and Costs for Two Policy Scenarios (Millions of Dollars)

	10% Discount Rate			*20% Discount Rate*		
	Benefits	*Costs*	*Net*[a]	*Benefits*	*Costs*	*Net*[a]
Scenario I:						
No Control	0.0	0.0	0.0	0.0	0.0	0.0
Sulfur Law	24.0	0.6	23.4	43.0	0.6	42.4
1975 Coal Ban	31.1	(0.1)	31.2	57.0	0.3	56.7
1975 Coal and Oil Ban	34.9	3.7	31.2	63.7	6.0	57.7
Scenario II:						
No Control	0.0	0.0	0.0	0.0	0.0	0.0
Sulfur Law	24.0	0.6	23.4	43.0	0.6	42.4
1977 Coal Ban	28.7	0.0	28.7	51.6	0.4	51.2
1977 Coal and Oil Ban	31.8	3.0	28.8	56.5	5.1	51.4

Note: The most likely assumptions apply. All values are annualized costs or benefits.
[a]Net = Benefits – Costs

considerable margin. In fact, under the basic assumptions fuel savings alone would justify the policy. Because of the uncertainties about the assumptions in the analysis, it is reasonable to ask under what circumstances the coal ban is not beneficial to society. The following assumptions are used in various combinations to help answer this question:

Endogenous price increases in natural gas:

1. The supply curve is infinitely elastic–i.e., there are no endogenous price increases;
2. The supply curve has an elasticity of 5; or
3. The supply curve has an elasticity of 3.

Exogenous oil and gas price increases:

1. No exogenous price increases occur;
2. The coal adjustment factor is .768; or
3. The coal adjustment factor is .406 (see Appendix G).

Conversion costs:

1. Conversion costs are as presented in Table 2-2; or
2. Conversion costs are double those presented in Table 2-2.

Abandonments:

1. There will be no abandonments; or
2. There will be massive premature abandonments.

Benefits:

1. The minimum benefit estimate is appropriate;
2. The mid-range benefit estimate is appropriate; or
3. The maximum benefit estimate is appropriate.

These assumptions pertain to the 5 parametric analyses conducted in Chapters 2 and 4. There are 144 possible combinations of these assumptions. We are only interested in the sign of the net benefits. Thus most of the 144 assumption sets do not have to be explicitly analyzed. For example, consider the basic cost and benefit estimates given in Table 6-1. Since a maximum cost estimate is made when using an elasticity of 3 and since the net benefits are positive, the net benefits would also be positive if either of the other two endogenous price assumptions are used. Similarly, net benefits would be positive if the highest benefit estimate is used rather than the mid-range estimates. Table 6-3 summarizes the parametric analyses and is used to determine the conditions under which the coal ban should not be implemented.

Because benefits are affected by exogenous price changes and abandonments, 6 assumptions sets, which produce maximum costs and minimum benefits, must be examined for each discount rate. For a 20 percent discount rate the net marginal benefits of the 1975 coal ban for these assumption sets are all positive. This implies that net marginal benefits are positive for all 144 possible assumption sets for this discount rate. For a 10 percent discount rate there exist 2 extreme assumption sets under which the coal ban is not justified. Calculations not shown reveal that within the ranges of the parametric analyses considered, endogenous and exogenous price changes do not affect these results. However, if conversion costs are actually closer to those presented in Table 2-2, if massive abandonments do not occur, or if the mid-range benefit estimates are more appropriate, then the net benefits would be positive. The latter two conditions are the most important.

The costs associated with massive abandonments assume that there are no alternative uses for abandoned property and that additional resources would have to be employed in the residential sector to supply additional dwelling services for the displaced families. If either of these two assumptions is relaxed, then the estimated social loss of $300 per dwelling unit per year would be reduced. Since these losses are more than 90 percent of the total social costs of abandonments, a relatively small reduction in the $300 figure would make the coal ban justified even under the extreme circumstances. Some people argue that there exists a surplus of low income housing in the city of Chicago. If this is true, then there would be no social costs associated with the abandoned dwellings, since no additional resources would have to be used to replace them.

Recall that the assumptions used to estimate the minimum value of the benefits are based on the property value approach and assign no benefits

Table 6–3. Parametric Analysis for the 1975 Coal Ban (Millions of Dollars)

	Assumptions					Marginal Annualized Values		
Case	*Supply Elasticity*	*Adjustment Factor*	*Conversion Costs*	*Premature Abandonments*	*Benefits*	*Benefits*	*Costs*	*Net Benefits (Costs)*
Discount Rate of 10 Percent								
1	3	None	Double Table 2–2	No	Minimum	2.4	0.0	2.4
2	∞	.406	Double Table 2–2	No	Minimum	4.5	0.7	3.8
3	3	.406	Double Table 2–2	No	Minimum	4.5	1.7	2.8
4	3	None	Double Table 2–2	Yes	Minimum	2.0	2.7	(0.7)
5	∞	.406	Double Table 2–2	Yes	Minimum	4.1	3.4	0.7
6	3	.406	Double Table 2–2	Yes	Minimum	4.1	4.3	(0.2)
Discount Rate Of 20 Percent								
1	3	None	Double Table 2–2	No	Minimum	4.5	1.5	3.0
2	∞	.406	Double Table 2–2	No	Minimum	8.2	2.7	5.5
3	3	.406	Double Table 2–2	No	Minimum	4.5	3.8	0.7
4	3	None	Double Table 2–2	Yes	Minimum	3.8	3.8	0.0
5	∞	.406	Double Table 2–2	Yes	Minimum	7.5	5.0	2.5
6	3	.406	Double Table 2–2	Yes	Minimum	7.5	6.1	1.4

Sources: Tables 2–7 through 2–10, Table F–9 and Table 4–11.

for sulfur dioxide improvements. If the mid-range benefits for particulates *or* the minimum positive benefits for sulfur dioxide are assumed, then the coal ban would be justified under all the parametric changes.

The 2 assumption sets that generate a negative net benefit for the 1975 coal ban are highly unlikely to occur. They are also based on assumptions that would produce positive net benefits if they were slightly altered. Therefore, it is concluded that the 1975 coal ban is the optimal residential fuel policy for the Chicago metropolitan area.

There are three additional results of interest related to total costs and benefits. First, if the minimum benefit valuation is appropriate, then the annual net benefits of the sulfur law would be \$.6 million and \$1.0 million, for 10 and 20 percent discount rates, respectively. This implies that the minimum benefit derived from reducing particulate levels is greater than the costs of the sulfur law even though the goal of the law was to reduce sulfur dioxide levels.

Secondly, in Chapter 2 it was mentioned that endogenous price increases would affect non-residential winter gas consumers and these costs are social costs of the policy. Furthermore, air quality improvements resulting from residential control also reduce damage to non-residential property. (If it were feasible, mortality, health and morbidity benefits should be estimated separately for work areas and residential areas and weighted by the time spent at work and at home.) Neither of these costs or benefits were included in the benefit-cost comparisons made previously. Based on the discussion in Appendix H, the value of the benefits to the non-residential sector for air quality improvements is about equal to the values of the material damage reduction benefits to residences. The proportion of the total benefits attributed to material damage reductions is about 50 percent (see Table 4-6). Using the minimum value of the marginal benefits of the 1975 coal ban implies additional benefits of \$1.2 million and \$2.2 million per year for 10 and 20 percent discount rates, respectively. The endogenous price increases would cost non-residential establishments about \$.65 million and \$.72 million annually for discount rates of 10 and 20 percent, respectively (this is equal to the costs to the residential sector, since about half of the winter gas to Chicago is consumed by non-residential activities). Since the sign of the net benefits of the 1975 coal ban remains the same for all the extreme cases in Table 6-3 when non-residential activities are included in the analysis, the optimality of the coal ban is not affected.

Finally, the space heating fuel policies proposed by the state of Illinois and the federal Environmental Protection Agency do not distinguish between residential and non-residential space heating sources. Because of data limitations, this study was restricted to the residential sector. The parameters that affect the results are source (stack) characteristics, fuel prices, conversion costs, and meteorological conditions. Since these parameters would be similar or identical for non-residential users and the same population would be benefited, it is reasonable to conclude that an independent study of non-residential establishments would produce positive marginal net benefits. If the endogenous price increase considering demands by both residential and non-residential users, would be similar to that calculated for the coal and oil ban, then the combined annual costs would increase by about $3 million. (This figure is an overestimate because there is less coal used for non-residential space heating than oil used for residential space heating.) Thus, extension of the coal ban to non-residential users would be justified except under the extreme assumptions that the minimum benefits apply and that many businesses would be abandoned.

The distribution of economic impacts of a coal ban was analyzed in the preceding chapter for geographic areas and economic sectors. The results indicated that a coal ban would not generate extreme inequities although some compensation policy could be justified. It is therefore reasonable to conclude that a 1975 ban on coal for space heating is an energy–environmental policy justifiable on economic grounds.

THE BENEFIT-COST APPROACH AND THE STANDARDS APPROACH

In Chapter 3 it was stated that none of the residential fuel policies could be identified as being optimal based on the standards approach. Assuming that the standards were properly determined, then the minimum cost solution for obtaining the standard would be the best. This is the premise of cost-effectiveness analysis. However, assuming that industrial and utility regulations are fixed, none of the residential policies would meet the annual federal secondary standard for particulate matter with any reasonable degree of confidence. In fact, the probability of meeting the standard only ranged between .25 for the no control policy and .32 for the coal and oil ban. Because there is no apparent policy that is superior, any one of them may be selected as a good policy. Even if the standard was met for one policy and not another, the small difference between the policies in terms of estimated impacts on air quality, suggests that the policy choice may be very sensitive to the assumptions used in the analysis—i.e., the result may not be very robust.

In addition to the inadequacies of the standards approach in selecting the best policy from various alternatives with similar effects on air quality, the appropriateness of the standards has been questioned. The recent abolishment of the annual secondary standard for sulfur oxides is a case in point. The standards approach may also motivate some degree of over-control, as discussed in Appendix H. Although the level of control may still provide positive net benefits, the additional benefit of the last degree of air quality improvement may be substantially outweighed by the additional costs.

This study has demonstrated that benefit–cost analysis of environmental policies in urban areas is feasible using presently existing analytical tools. The benefit–cost analysis in this study was able to identify the 1975 coal ban as a superior residential fuel policy. In addition, this was done without a need to consider standards. Since the benefit–cost approach has advantages over the standards approach, it is important to identify areas of additional research that will improve the accuracy of the analysis.

One possible direction for future research is to better quantify morbidity and material damage effects. This study has suggested that these benefits considerably outweigh the benefits derived from extending the length of people's lives. Recent research has concentrated heavily on isolating the effects of air pollution on mortality rates. Additional research is needed to isolate the effects of peak pollutant concentrations on a seasonal, daily, and hourly basis. This would have been a great help in the present study, since the major effects of air pollution generated by space heating occur in winter.

An additional problem of the benefit–cost approach associated with large improvements in air quality results from the use of linear coefficients in most benefit estimation. If the total damage function is not linear (e.g., convex, or concave), damage reductions may be inaccurately estimated. (See Appendix H for a more detailed discussion of this problem). In addition, threshold values for certain deterioration effects may exist. These difficulties are less important when only a small improvement in air quality is considered, such as a residential fuel policy. To evaluate a complete environmental program, additional research must be conducted and new techniques developed.

Improvements in the predictive capability of atmospheric dispersion models for annual, seasonal, daily, and hourly air quality levels are necessary to improve both the standards and benefit–cost approaches. Some of the major problems with the AQDM model that was used in this study and is sanctioned by the federal Environmental Protection Agency are: the virtual point source is used for simulating the effects of area source emissions; persistent levels of meteorological conditions are not considered; all particulate matter is treated as a gas, ignoring heavy particles; the influence of moisture, temperature, and other parameters that affect air quality levels are not considered; and a linear calibration curve is used.

Finally, it has been argued that the value of human life cannot be quantified, and that premature deaths related to pollution are intolerable. This implies that the willingness to pay approach used in this study is inappropriate for evaluating the benefits of air pollution improvements. Indeed, this is in part the argument used to justify the standards approach. One can not quarrel with the concept that a human life may be priceless; however, the incompleteness of the argument should be noted. If a pollution standard is set solely on the the basis that deleterious effects of the pollutant exist, it is quite possible that the adverse effects of meeting the standard would be worse. Consider the possibility that pollution controls may increase the frequency of electricity blackouts or fuel shortages. The number of people who may be killed by automobiles on darkened streets, the number of people who die or become seriously ill from living in inadequately heated buildings, or the number of people who die because emergency care or vehicles are not available may be considerably greater than the deaths or illnesses caused by pollution. In a benefit–cost framework, these adverse effects of pollution control would be considered as costs (or negative benefits) of the policy.

IMPLICATIONS FOR NATIONAL ENERGY AND ENVIRONMENTAL POLICIES

Implementing a ban on the use of coal for space heating, which was found to be the most beneficial policy in this study, would result in additional annual demands for natural gas and oil. Although the quantities of fuel required represent only a small fraction of the total consumption of natural gas and oil in Chicago, the relationship between these demands and energy supplies must be discussed in light of energy problems in the United States

In the absence of price controls, excess demands can not exist because prices would increase over time driving quantities supplied upward or quantities demanded downward. In this analysis two types of price increases for natural gas were considered. It was observed that an increase in the wellhead price of natural gas is required to provide sufficient supplies of gas nationally. In this study, these were called exogenous price increases. Local endogenous price increases would eliminate possible local gas shortages resulting directly from the coal ban.

Even with these price increases, it was still assumed that there might be a shortage of natural gas in Chicago on a short-term basis, which would force some coal users to convert to oil. Since conversion costs, operating and maintenance costs, and emissions are greater for oil conversions than gas conversions, this assumption overestimates the costs and underestimates the benefits of the coal ban. Therefore, if sufficient supplies of natural gas are available in Chicago to convert more coal users to gas, the coal ban becomes more attractive.

The fact that fuel restrictions continue to appear justified when substantial price increases are projected for clean fuels gives confidence in the conclusion that similar restrictions on the use of coal for all space heating in Chicago and elsewhere would be beneficial. The following additional considerations support this conclusion:

1. Although uncontrolled particulate emissions per Btu of coal consumed are greater for non-space heating sources, the many control devices available for controlling particulate matter are only applicable to industrial and utility sources. Therefore, particulate emissions per Btu of coal consumed are greater for space heating sources.

2. Since space heating furnaces tend to be less efficient than industrial boilers, they require more coal per Btu of output. This implies that space heating generates more sulfur dioxide per Btu than other fuel combustion activities. In addition, there is an emerging technology for controlling sulfur dioxide emissions from utility and industrial sources.
3. The effective height of emission release—i.e., stack height plus plume rise—is, in general, less for space heating sources. Therefore, the impact on ground level air quality is greater per ton of pollutant released into the atmosphere for space heating sources.

These facts imply that air quality benefits per Btu are greatest for policies requiring the use of clean fuels by space heating sources. Because there are alternatives to fuel switching for industrial and utility sources—i.e., feasible control devices are becoming available—environmental benefits outside the residential sector can be realized without switching to cleaner fuels.

The firmest conclusions of this study pertain to coal. Every indication is that coal will continue to decline as a household fuel, even if prices of alternative fuels rise. The substantial environmental cost of using coal in the home strengthens the argument that coal as an energy source should be used elsewhere.

It should be emphasized that empirical results in this study were obtained only for Chicago. Further analysis of policies, applying the methods developed here, is needed in other cities.

Appendix A

Chicago Area Air Quality Regulations

The 1970 amendments to the Clean Air Act (42 U.S.C. 1857) required each state to submit implementation plans for obtaining and maintaining federal air quality standards, including sulfur dioxide and particulate matter standards. The implementation plans include regulations for limiting sulfur dioxide and particulate emissions from fuel combustion and process sources, and particulate emissions from incinerators.

The city of Chicago passed its Environmental Control Ordinance prior to any state action. Of particular importance is the regulation restricting the sulfur content of fuels sold or used in Chicago. Table A-1 summarizes these limits as they pertain to the period 1968 through 1972.

The regulations that were adopted by the states of Illinois and Indiana were simulated in the control regulation model discussed in Chapter 3. The regulations modeled for existing and new sources are presented in this appendix. (New source performance standards were promulgated by the federal government and each state's plan had to be consistent with these standards.[1]) Various exceptions to these regulations provided in each state's plan were not modeled. However, some amendments to the original regulations were considered to be applicable. The description of each regulation in this appendix is preceded

Table A-1. Chicago Sulfur Law Compliance Schedule (Percent Sulfur Limit by Weight)

Heating Season	*Industry*	*Utility*	*Residential/Commercial*
1968-69	3.00[a]	3.00	3.00
1969-70	2.50	2.50	2.50
1970-71	1.50	1.80	2.00
1971-72	1.50	1.80	1.25
1972-73	1.00	1.00	1.00

[a]A three percent sulfur limit is assumed when no limits are specified.

by a number which refers to the Illinois Pollution Control Board rules or the Indiana Air Pollution Control Board Regulations.

ILLINOIS SULFUR DIOXIDE REGULATIONS AS MODELED

Fuel Combustion: (Rule 204.a-d)

No person shall cause or allow the emission of sulfur dioxide into the atmosphere in any one hour period from any fuel combustion emission source to exceed:

$$E = S_c H_c + S_r H_r + 0.3 H_d + 0.0034 H_g$$

where

E ≡ allowable sulfur dioxide emission rate in pounds per hour;

S_c ≡ coal sulfur dioxide emission standard in pounds per million Btu;

S_r ≡ residual oil sulfur dioxide emission standard, in pounds per million Btu, which is applicable;

H_c ≡ actual heat input from coal in million Btu per hour;

H_r ≡ actual heat input from residual oil in million Btu per hour;

H_d ≡ actual heat input from distillate oil in million Btu per hour;

H_g ≡ actual heat input from natural gas in million Btu per hour;

H ≡ total heat input in million Btu per hour–i.e., $H = H_c + H_r + H_d + H_g$

For all existing sources, new sources with $H \leqslant 250$, $S_c = 1.8$ and $S_r = 1.0$. For new sources with $H > 250$, $S_c = 1.2$ and $S_r = 0.8$.

Process: (Rule 204.f)

No person shall cause or allow the emission of sulfur dioxide into the atmosphere from any process emission source to exceed 2000 ppm.

INDIANA SULFUR DIOXIDE REGULATIONS AS MODELED

Fuel Combustion: (Regulation APC 13)

Maximum total sulfur dioxide emissions from existing fuel-burning operations and new-emission sources with a heat input less than or equal to 250 million Btu/hr are limited to:

$$E = S H$$

$$S = 17.0\, Q^{-0.33}$$

subject to

$$1.2 \leqslant S \leqslant 6.0$$

where

E ≡ allowable emission rate in pounds per hour;

S ≡ allowable sulfur dioxide in the stack gases in pounds per million Btu;

Q ≡ rated capacity of heating equipment in million Btu per hour;

H ≡ actual heat input in million Btu per hour.

Maximum total sulfur dioxide emissions from new fuel-burning operations with heat input more than 250 million Btu/hr are limited to

$$E = 1.2\, H,$$

where E and H are defined above.

Process: (Regulation APC 13)

Maximum total sulfur dioxide emissions from all process operations shall be limited to:

$$E = 19.5\, P^{0.67}$$

where

E ≡ allowable emissions in pounds per hour;

P ≡ process weight in tons per hour.

ILLINOIS PARTICULATE REGULATIONS AS MODELED

Fuel Combustion (Rule 203.g)

No person shall cause or allow the emission of particulate matter into the atmosphere in any one hour period from any new or existing source to exceed:

$E = .1\,H_c + .1\,H_\ell + .017\,H_g$

where

E ≡ allowable particulate emission rate in pounds per hour;

H_c ≡ actual heat input from coal in million Btu per hour;

H_ℓ ≡ actual heat input from oil in million Btu per hour;

H_g ≡ actual heat input from natural gas in million Btu per hour.

Process: (Rule 203.a)

No person shall cause or allow the emission of particulate matter into the atmosphere in any one hour period from any existing process emission source, either alone or in combination with the emission of particulate matter from all other similar new or existing process emission sources at a plant or premises, to exceed:

$E = 4.10\,P^{0.67},\ P \leqslant 30$

$E = [55.0\,P^{0.11}] - 40.0,\ P > 30$

where

E ≡ allowable emission rate in pounds per hour;

P ≡ process weight rate in tons per hour.

No person shall cause or allow the emission of particulate matter into the atmosphere in any one hour period from any new process emission source, either alone or in combination with the emission of particulate matter from all other similar new process emission sources at a plant or premises, to exceed:

$E = 2.54\,P^{0.534},\ P \leqslant 450$

$E = 24.8\,P^{0.16},\ P > 450,$

where

E ≡ allowable emission rate in pounds per hour;

P ≡ process weight rate in tons per hour.

Incineration: (Rule 203.e)

No person shall cause or allow the emission of particulate matter into the atmosphere from any incinerator to exceed:

$E = 0.05\,(1.42 \times 10^{-4})\,(V)$, for new and existing sources with $P > 60{,}000$;

$E = 0.08\,(1.42 \times 10^{-4})\,(V)$, for new and existing sources with $2{,}000 < P \leqslant 60{,}000$;

$E = 0.2\,(1.42 \times 10^{-4})\,(V)$, for existing sources with $P \leqslant 2{,}000$; or

$E = 0.1\,(1.42 \times 10^{-4})\,(V)$, for new sources with $P \leqslant 2{,}000$;

where

$E \equiv$ allowable emission rate in pounds per hour;

$P \equiv$ process weight rate in pounds per hour;

$V \equiv$ standard cubic feet of effluent gases per hour–i.e., $V = A\,(273/T)$;

$A \equiv$ actual cubic feet of effluent gases per hour;

$T \equiv$ stack gas temperature in degrees Kelvin.

INDIANA PARTICULATE REGULATIONS AS MODELED

Fuel Combustion: (Regulation APC 4-R)

The emission of particulate matter from the combustion of fuel for indirect heating shall be limited to:

$E = Pt\,H$

$Pt = 0.87\,Q^{-0.16}$

subject to

$0.2 \leqslant Pt \leqslant 0.6$,

where

$E \equiv$ allowable emissions in pounds per hour;

$Pt \equiv$ pounds of particulate matter emitted per million Btu heat input;

$H \equiv$ actual heat input in millions of Btu per hour;

$Q \equiv$ total plant operating capacity rating in million Btu heat input per hour.

Emissions of particulates from the combustion of fuel in new stationary installation for indirect heating in excess of 250 million Btu per hour heat input shall be limited to

$$E = 0.1\,H,$$

where E and H are defined above.

Process: (Regulation APC 5)

No person shall operate any process so as to produce, cause, suffer, or allow particulate matter to be emitted in excess of:

$$E = 4.10\,P^{0.67},\ P \leqslant 30$$

$$E = [55.0\,P^{0.11}] - 40.0,\ P > 30,$$

where

$E \equiv$ allowable emission rate in pounds per hour;

$P \equiv$ process weight rate in tons per hour.

Incineration: (Regulation APC 6)

No person shall cause or permit the emission of particulate matter from the stack or chimney of any incinerator in excess of:

$$E = 0.2\,(1.42 \times 10^{-4})\,(V), \text{ for existing sources with } P > 1{,}000;$$

$$E = 0.35\,(1.42 \times 10^{-4})\,(V), \text{ for existing sources with } P \leqslant 1{,}000; \text{ or}$$

$$E = 0.1\,(1.42 \times 10^{-4})\,(V) \text{ for all new sources;}$$

where

$E \equiv$ allowable emission rate in pounds per hour;

$P \equiv$ process weight rate in pounds per hour;

$V \equiv$ standard cubic feet of effluent gases per hour–i.e., $V = A\,(273/T)$;

$A \equiv$ actual cubic feet of effluent gases per hour;

$T \equiv$ stack gas temperature in degrees Kelvin.

NOTES TO APPENDIX A

1. U.S. Environmental Protection Agency, *Background Information for Proposed New-Source Performance Standards* (Research Triangle Park, N.C., 1971).

Appendix B

The Coal User Survey

During the course of our study, we decided that various hypotheses about coal users had to be tested. In particular, we assumed that the accelerated trend away from coal after 1968 would continue and that it primarily resulted from coal price increases. Furthermore, the extent of abandonment or demolition, an important consideration in the parametric analysis, depends on the age distribution of the buildings, for which there were no available data. Finally, the available information on fuel usage and prices needed to be substantiated. Therefore, a survey of coal users in Chicago was conducted.

This appendix summarizes the major findings of the survey and describes the sampling procedure used. This procedure was developed by Dr. Thomas Baldwin of Argonne National Laboratory and the survey was conducted under his supervision.

RESULTS AND CONCLUSIONS

Out of 200 telephone calls, 61 interviews were completed. Forty of the buildings were still using coal, while 21 had already converted from coal. Figure B-1 shows the age distribution of the buildings. One finds that 58 percent of the existing coal users and 39 percent of those buildings already converted are over 49 years old. This suggests that conversion is less attractive for older buildings and the hypothesis that there is a higher probability of abandonment of older buildings is reasonable. Furthermore, the age distribution of coal-burning buildings is heavily weighted toward these older buildings.

Other results obtained from the survey are:

1. The age distribution of the boilers is skewed toward the older boilers. This supports our hypothesis that these boilers are approaching the end of their economic life.

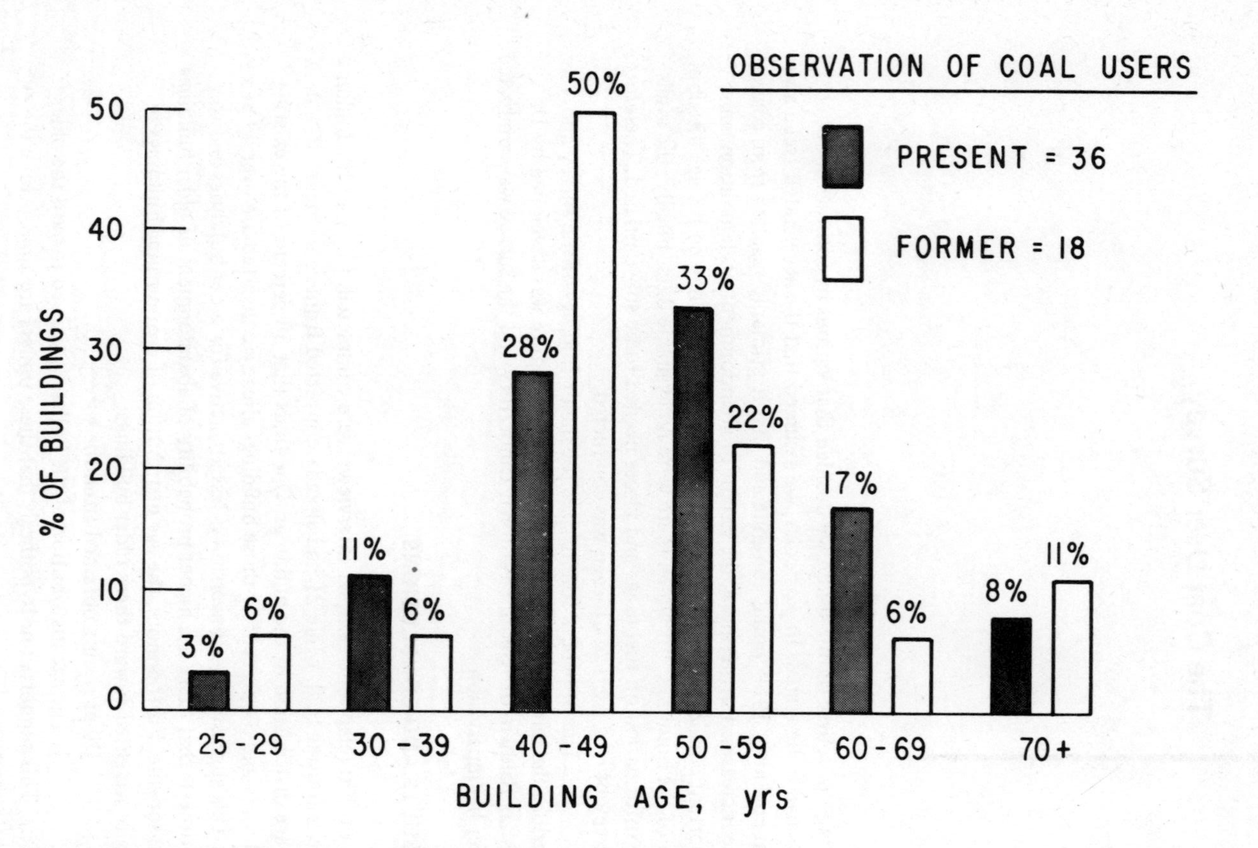

Figure B–1. Age Distribution of Coal-Using Buildings, Chicago, 1972.

2. Twenty-five percent of the buildings that presently use coal have an occupancy rate less than or equal to 90 percent, while only 13 percent of those who already converted have such a low occupancy rate. This suggests that conversion from coal might be more attractive in buildings with low vacancy rates or that conversions increase occupancy. Furthermore, the relationship between the decision to convert the boiler or abandon the building is affected by the vacancy rate as discussed in Appendix D.

3. The range of prices paid for a ton of coal in the 1972 to 1973 heating season is much wider than the range of prices quoted by coal distributors for low sulfur coal. Possibly some of these buildings are using a higher sulfur coal, and others may be dealing with distributors who are making windfall profits. In general, however, the prices quoted by coal distributors are consistent with the prices provided by the coal users.

4. Two-thirds of the buildings surveyed are residential buildings. This suggests that the results of the survey are representative of all coal-using residential buildings in Chicago.

Figure 1-4 summarizes the most important result of the survey, the reasons people are making conversions. Coal users were asked, "Have you considered converting, and if so, why?" Former coal users were asked, "Why did you convert to your present fuel?" In all, 108 reasons were given for converting from coal. Over 50 percent of the answers from those who had already converted and those who plan to convert emphasized economic factors as the motivation for conversion. We aggregated the reasons of cheaper, rising costs, less janitorial labor, less boiler maintenance and more efficient heat as economic reasons.

Because of the open-ended nature of the questionnaire, we were unable to ascertain how much of an influence the low sulfur law had in making a decision to convert. Certainly, the related price increase in coal and the above result that conversions are highly sensitive to fuel costs suggest that the sulfur law had a significant role in increasing the rate of conversions. Furthermore, a large number of the respondents stated that pollution control regulations influenced their decision. Specifically, the two most frequent responses were the existing sulfur law and the proposed coal ban.

Another result confirms our assumption that the accelerated trend will continue if the relative prices of coal to oil and gas remain the same–that is, many of the existing coal users stated that they have not converted yet because they were placed on a waiting list for natural gas. As soon as the gas is made available, they will make the conversion. This also gives weight to our assumption that some people will not be able to get gas under a coal ban and would have to convert to oil.

SAMPLING PROCEDURE

Subjects for the "Coal Users Survey" were sampled from the city of Chicago's boiler inspection inventory.[a] This inventory locates all coal-heated buildings in square mile cells of a grid superimposed on a map of the city. The buildings in each cell are listed by a sequence number and the name and address of the owner/agent are given.

Preliminary examination of frequency distributions revealed that coal-heated buildings were unequally distributed over the city. Therefore, in order to obtain a reasonable representation of the 8276 buildings inventoried, a proportional stratified sample of 200 was drawn.

The square mile cells were first classified into seven strata, depending upon the number of coal-heated buildings they contained. Figure B-2 shows these strata and indicates that buildings heated with coal are clustered in the northeastern and southeastern sectors of the city. Next it was determined what proportion of all the city's coal-heated buildings were located in each square mile. This proportion was then multiplied by 200 so that the number of buildings sampled could be apportioned over the various square miles. The number obtained at this point was rounded to the nearest whole number to indicate how many buildings would be drawn from each cell in the grid. Because only 16 buildings were required from the 105 cells containing 1 to 24 coal-heated buildings, 16 square-mile cells were visually picked at random and 1 building was sampled from each of these areas.

Having stratified the square miles and determined the proportionate number of cases to be selected from each, specific buildings were drawn using the table of random numbers. A starting point was randomly selected on the table of random numbers. Then, beginning with the square miles with the highest concentrations of coal-heated buildings–i.e., 200 or more–starting points were randomly selected on each list of buildings. These lists were then counted through using sequential numbers in the table of random numbers and buildings were extracted until the prescribed number had been sampled from all square miles in the inventory. The addresses of these sampled buildings were then listed along with the owner or agent's name and address. Finally, telephone numbers for the owner or agent were obtained from the Chicago and suburban directories or from the telephone company's information services. These persons were telephoned and asked to respond to the survey.

Of the 200 buildings surveyed, only 61 interviews were completed. It was impossible to complete 139 interviews for the following reasons:

Owner or agent refused to answer questionnaire	20
Owner or agent telephone number could not be obtained	57

[a]We are indebted to the Chicago Department of Environmental Control for providing this inventory updated to December 21, 1972.

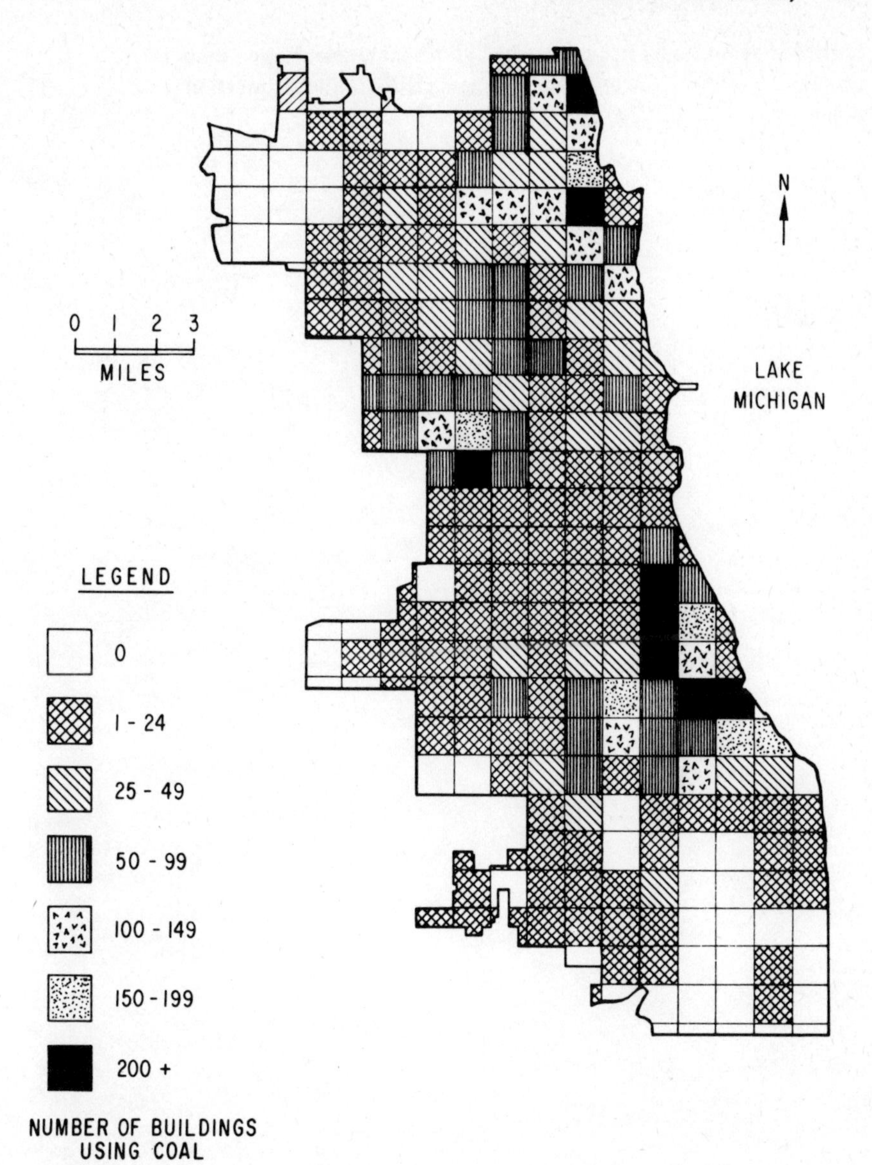

Figure B–2. Spatial Distribution of Coal-Using Buildings, Chicago, 1972.

Owner or agent could not be reached at available telephone number	29
Owner or agent no longer owned or managed building in question	17
Other	15
	139

Appendix C

Engineering Cost Model for Conversion and Fuel Usage

An engineering-cost model was developed for this study by Dr. Lyndon Babcock and Mr. Niren L. Nagda of the University of Illinois School of Public Health in Chicago. The model estimates the cost of switching from coal to oil, gas, or electricity, and the annual operating and maintenance costs of using each fuel for residential space heating. These costs are calculated for buildings ranging in size from single-family homes to large apartment buildings with up to 150 dwelling units.

The main independent variable used in the program is the number of dwelling units in a building. Output-heat-rate requirements of a furnace are calculated by using engineering estimates of the relationship between energy consumption and the number of dwelling units in a building. The heat-rate requirements are then used to calculate the cost of the basic heating equipment needed to convert the furnace (e.g., the burner) and accessories such as a gas line or oil tank. The total capital cost, which is the sum of the equipment and installation costs, is calculated. Yearly fuel usage is estimated using the output-heat-rate requirements, efficiency of the furnace, heat value of the fuel, and load factor—i.e., utilization factor—of the furnace. Monthly fuel costs are calculated using fuel price schedules and estimated monthly fuel consumption or purchasing patterns. Thus, the effects of quantity discounts are simulated. Labor costs are estimated on the basis of the size of the furnace and the type of fuel used.

Costs of converting to oil, gas, or electricity, are annualized over the expected life of the furnace. These costs are added to the associated annual operating and maintenance costs, for each of these fuels. The model compares these totals with the cost of heating with coal. Conversion costs and operating and maintenance costs are calculated for different years by using historical fuel price data and various economic indices to adjust labor costs and to account for inflation.

Table C-1 shows the input and calculated variables used in the program. The first column indicates values of the variables when the coal heating

Table C-1. Input and Calculated Program Variables

			COAL	*GAS*	*OIL*	*ELECTRIC*
1	INSTALLATION YEAR		1930.000	1970.000	1970.000	1970.000
2	SALVAGE YEAR		1970.000	1990.000	1990.000	1990.000
3	CAPITAL COST	($)	381.000	1513.554	3679.471	7136.277
4	DISCOUNT RATE		0.060	0.080	0.080	0.080
5	FUEL COST ($/UNIT)		19.306	0.083	0.162	0.013
6	EFFICIENCY OF BOILER		0.567	0.800	0.750	0.950
7	HEAT RATE OUTPUT (BTU/HR)		892035.125	892035.125	892035.125	892035.125
8	OPERATING LABOR (HRS/YR)		699.114	20.000	40.000	5.000
9	OPERATING LABOR COST	($/YR)	1900.034	54.355	108.711	13.589
10	SALVAGE VALUE	($)	19.050	75.678	183.974	356.814
11	LOAD FACTOR OF BOILER		0.300	0.300	0.300	0.300
12	BTU OUTPUT PER UNIT OF FUEL		26000000.0	100000.000	139000.000	3413.000
13	DWELLING UNITS PER BUILDING		34.000	34.000	34.000	34.000
14	OPERATING LABOR WAGE	($/HR)	2.718	2.718	2.718	2.718
15	INSTALLATION LABOR WAGE	($/HR)	0.0	5.436	5.436	0.0
16	INSTALLATION LABOR COST	($)	0.0	217.422	217.422	0.0
17	COST OF CONVERSION PERMITS	($)	0.0	145.000	145.000	0.0
18	COST OF FUEL TANK AND LINE	($)	0.0	506.917	2672.833	0.0
19	COST OF BASIC EQUIPMENT	($)	0.0	644.216	644.216	0.0
20	FUEL COST ($/MBTU)		0.743	0.830	1.165	3.674
21	OPERATING LIFE (YEARS)		40.000	20.000	20.000	20.000
22	ANNUALIZED CAPITAL COST	($/YR)	25.199	152.520	370.764	719.068
23	HEAT INPUT (MBTU/YR)		4136.941	2930.334	3125.690	2467.650
24	FUEL QUANTITY (UNITS/YR)		159.113	29303.344	22486.980	723015.000
25	FUEL COST ($/MBTU)		1.310	1.037	1.554	3.867
26	TOTAL COST PER YEAR	($)	4997.066	2638.511	4122.359	9798.996
27	ADJUSTED COAL COST PER YEAR	($)	5029.574	0.0	0.0	0.0
28	COST INCREASE	($/YR)	0.0	-2391.063	-907.215	4769.422
29	COST INCREASE (%)		0.0	-47.540	-18.038	94.828
30	OIL TANK CAPACITY (GAL)		0.0	0.0	3000.000	0.0
31	FUEL USED IN MAY		5.410	996.314	764.557	24582.508
32	JUN FUEL AMT		1.273	234.427	179.896	5784.117

33	JUL FUEL AMT		0.0	0.0	0.0	0.0
34	AUG FUEL AMT		0.0	0.0	0.0	0.0
35	SEP FUEL AMT		2.068	380.943	292.331	9399.191
36	OCT FUEL AMT		8.433	1553.077	1191.810	38319.793
37	NOV FUEL AMT		19.412	3575.007	2743.411	88207.812
38	DEC FUEL AMT		28.799	5303.902	4070.143	130865.687
39	JAN FUEL AMT		31.186	5743.453	4407.445	141710.875
40	FEB FUEL AMT		27.049	4981.566	3822.787	122912.500
41	MAR FUEL AMT		23.071	4248.980	3260.612	104837.125
42	FUEL USED IN APRIL		12.411	2285.661	1753.985	56395.176
43	MAY FUEL COST	($)	104.442	88.675	123.858	311.649
44	JUNE FUEL COST, $		24.575	24.842	29.143	77.343
45	JUL FUEL COST, $		0.0	1.314	0.0	1.031
46	AUG FUEL COST, $		0.0	1.314	0.0	1.031
47	SEP FUEL COST, $		39.934	38.368	47.358	122.402
48	OCT FUEL COST, $		162.807	134.190	193.073	482.874
49	NOV FUEL COST, $		374.764	299.483	444.432	1104.688
50	DEC FUEL COST, $		556.002	429.485	659.363	1636.385
51	JAN FUEL COST, $		602.080	461.625	714.006	1771.561
52	FEB FUEL COST, $		522.212	405.915	619.291	1537.254
53	MAR FUEL COST, $		445.416	352.346	528.219	1311.959
54	APRIL FUEL COST	($)	239.603	194.079	284.145	708.169
55	ANNUAL FUEL COST	($/YR)	3071.836	2431.635	3642.888	9066.340

system is not converted. The other three columns indicate the values when a conversion is made in the "installation year"–i.e., row 1–to gas, oil or electricity, respectively.

The computer program is written in modular form and consists of a main program and 16 subroutines. The purpose of each subroutine is outlined briefly below. Subroutines are listed in alphabetical order. A flow diagram of the subroutines is presented in Figure C-1.

CALACC: Calculates the cost of accessories–i.e., oil tank, oil or gas lines, permits, and labor charges for installation.

CALANN: Calculates the annualized capital costs over the expected life of the heating equipment.

CALBTU: Calculates the output-heat-rate requirements from a relation between energy consumption and the number of dwelling units in a building.

CALCAP: Calculates the total capital cost of installation. This is the sum of the furnace, accessories, and labor costs. Appropriate economic indices are used to adjust these costs to specific years.

CALEQP: Calculates the cost of basic equipment as a function of the output-heat-rate required.

CALFUL: Calculates fuel costs using detailed fuel price schedules and monthly fuel consumption or purchase estimates. Annual fuel consumption is estimated from the output-heat-rate requirements; efficiency and load factor of the furnace; and heat value of the fuel.

CALOPR: Calculates the operating labor costs for each fuel by using the fuel consumption estimates and a labor cost index.

CALTOT: Aggregates the annualized capital, operational labor, and fuel costs.

CHANGE: Changes variables incrementally to generate data for plotting graphs.

DATAR: Reads input data.

FULPRC: Calculates fuel price schedules for each level of fuel usage.

GRAPHA: Plots graphs of the results.

PRINDX: Supplies the heating equipment index, labor cost index, and consumer price index.

RITEA: Writes out the data and results.

RITEB: Writes out the fuel price schedule.

ROUND: Rounds off the graph limits.

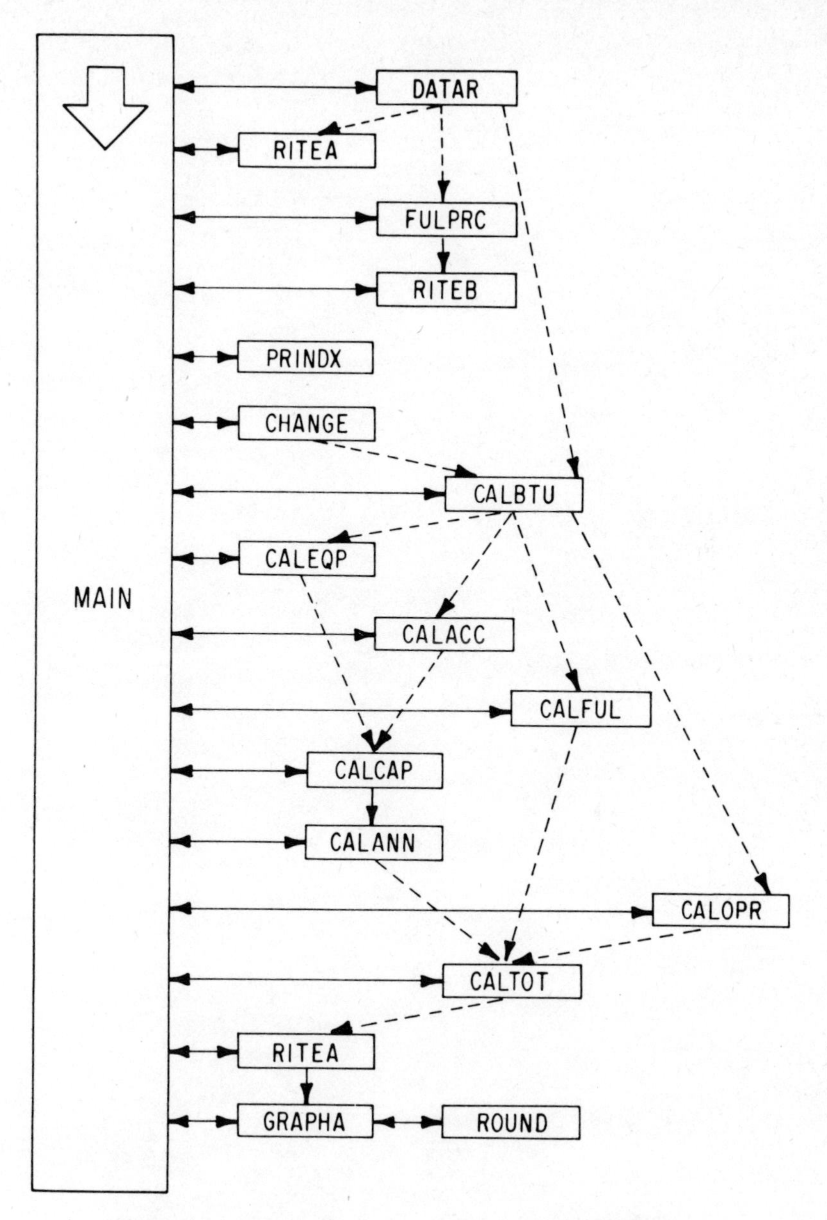

Figure C-1. Flow Diagram of Engineering Cost Model.

Appendix D

Relationships Between Rents, Conversion Costs and Abandonments in Chicago

Space heating fuel data indicate that the number of coal users in single-family homes and in duplexes is very small. These two building sizes have almost a one to one correspondence between occupants and owners. In the larger sizes, the percentage of owner-occupied buildings is still very small in spite of a recent trend toward condominiums. Thus, in general, space heating fuel policies affect landlords directly and tenants indirectly. In most buildings in Chicago, payments for space heating are included in the monthly rent. Hence, any increase in space heating costs would be borne by the landlords if rents are stable, or by the tenants when rents can be increased. In the following, we examine the relationship between rents and fuel conversion costs, and derive conclusions regarding the relationship between building abandonment and the rent paid.

The previous rent paid less operating expenses, for an apartment in a building that is abandoned because of the fuel policy, is part of the social cost of the fuel policy since it leaves real resources that would produce dwelling services of some positive value unused. (This is true only if the system is in equilibrium and if there is no alternative use of the property abandoned.) From a sectoral point of view these costs are imposed upon the landlords of the abandoned buildings. Furthermore, the displaced tenants incur moving expenses as well as possible inconvenience and psychic costs. Our working assumptions are that the population to be relocated can find alternative dwellings in nearby neighborhoods, and new buildings with clean fuels replace the abandoned buildings. Furthermore, it is assumed that the families displaced incur moving costs of $100 and inconvenience and psychic costs totaling $100 per year for the first 3 years after they were forced to move. This latter value is about 10 percent of their estimated annual gross rent. The 3-year estimate is the median number of years tenants have lived in their present locations in Chicago.[1]

Two sources of rent data are available for Chicago. The first is data published by the Institute of Real Estate Management.[2] The second is the Census of Housing.[3] The first source provides 1969 data; the second, 1970 data. These data are presented in Tables D-1 and D-2. The Real Estate Management data are *averages* that represent all buildings, not just those that might be abandoned because of a residential fuel policy. Thus, the reported average rents probably overestimate the rent figures which would be appropriate for our cost calculations.

In the two size categories represented in Table D-1, the majority of the buildings in the sample were built in the 1921 to 1930 period–i.e., they are now 40 to 50 years old. The average number of rooms per apartment is 3.5 to 4.0 and the net annual operating income $448 to $476. The share of operating expenses of the total gross rent ranges from 51 to 68 percent in low rise buildings of 12 to 24 units, and from 60 to 69 percent in low rise buildings of 25 units and over. In elevator buildings (not shown in the table) the range is from 49 to 73 percent. No relationship was found between these percentages and building age, when holding building size constant. A real estate manager claimed that the proportion of expenses in rent receipts increases as rent declines. Hence, for low rent apartments, the share of expenses in gross rent is assumed to be 65 percent.

Using the Peoples Gas, Light and Coke Company housing inventory data, abandonment rates by fuel were calculated that are believed to be within an order of magnitude of the actual rates. The rates of abandonment of coal-using buildings were calculated to be 3 percent and 5 percent in the 1968 to 1971 period, assuming that 30 and 50 percent of all abandonments are in coal using buildings, respectively. Similarly, assuming that 30 and 50 percent of all abandonments took place in oil heated buildings, abandonment rates would be 4 percent and 7 percent, respectively in the 1968 to 1971 period. Therefore, a 5 percent "natural" abandonment rate is likely to be close to the actual rate for both coal-and oil-using buildings with a sulfur law.

Using this 5 percent abandonment rate, an estimated 66 and 64 thousand coal- and oil-using dwelling units, respectively, will be abandoned over the 1970 to 1990 period. They represent 17.7 percent of the rented living units in 1970–i.e., (100) (64 + 66)/735 = 17.7. If it is assumed that these units are all in the lower rent categories shown in Table D-2, then the maximum rent paid is between $80 and $99 per month. Using the mean bracketed rent values and $25 for the lowest rent category, the estimated average gross rent paid is $71.3 per month or $856 per year. Based on the estimated 65 percent operating expense estimate from the Real Estate Institute data this implies an average net rent of $300 per year ((1 - .65) 856 = 299.6). This figure is consistent with the average net revenue per apartment reported in Table D-1 for buildings built between 1921 and 1930. Since it is likely that the buildings to be abandoned are a subset of this group, the $300 figure is used in the analysis to

Table D-1. Rent Data for Chicago, 1969

	Year Built				
	1961 to date	*1946–1960*	*1931–1945*	*1921–1930*	*1920 and before*
I. Low Rise Buildings of 12 to 24 Units					
No. of Buildings	13	18	5	59	6
No. of Apartments	208	293	89	1024	108
No. of Rentable Rooms	888	1274	294	4068	480
Total Actual Collections per Room per Annum ($)	506.7	458.1	341.4	366.5	378.3
Total Expenses per Room per Annum ($)	264.3	287.2	244.7	253.5	218.1
Net Operating Income per Room per Annum ($)	242.4	170.9	96.7	113.0	160.2
Net Operating Income per Dwelling Unit per Annum ($)	1034.9	743.1	319.4	448.9	712.0
II. Low Rise Buildings of 25 Units and Over					
No. of Buildings	6	33	5	66	7
No. of Apartments	312	2022	225	2929	269
No. of Rentable Rooms	1263	9386	1029	10514	1002
Total Actual Collections per Room per Annum ($)	370.8	446.2	366.9	376.8	316.2
Total Expenses per Room per Annum ($)	244.4	277.6	223.9	244.0	218.5
Net Operating Income per Room per Annum ($)	126.4	168.6	143.0	132.8	97.7
Net Operating Income per Dwelling Unit per Annum ($)	511.7	782.6	654.0	476.7	363.9

Source: Institute of Real Estate Management, *Apartment Building Income–Expense Analysis 1970* (Chicago: National Association of Real Estate Boards, 1970).

Table D-2. Distribution of Gross Rent, Chicago, 1970

Rent per Month	*Number of Families*	*Percentile*
Less than $30	799	0.1
30–39	2,359	0.4
40–49	9,192	1.7
50–59	16,371	3.9
60–69	26,988	7.6
70–79	39,043	12.9
80–99	116,476	28.8
100–119	151,524	49.4
120–149	198,924	76.5
150–199	125,625	93.6
200–249	26,964	97.2
250–299	9,293	98.5
Over 300	11,011	100.0
Total	734,569	
Median Rent	121	

Source: U.S. Bureau of the Census, *1970 Census of Housing,* Vol. 1, Part 15 (Illinois) (Washington: U.S. Government Printing Office, 1972), p. 195.

estimate the value of the dwelling services lost due to premature abandonment resulting from a fuel policy.

The maximum cost per dwelling unit for converting from coal to oil for buildings with more than two apartments is $450 and to gas is $383. (Conversion cost figures are from Table 2-2.) These costs decline as the size of the building increases. If a conversion is made, the landlord would reduce his fuel costs or janitorial costs as discussed in Chapter 2. Even if these savings are ignored, a conversion would be preferred to abandonment if net rents were $300 per year per apartment since the conversions could be paid for in less than two years out of these net revenues. (This assumes the value of the property to the landlord if the building is abandoned is zero; if the residual value of the property is positive, a longer life span is needed to justify conversions.) Therefore, as suspected, the requirement to convert a furnace is not a reason in itself to abandon a building. The buildings that are abandoned would probably require additional upgrading, have a high vacancy rate,[a] have a short life expectancy or have some other negative amenities that make the landlord decide to abandon rather than convert. In this study it is assumed that all the buildings that would have been abandoned over the 1975 (1977) to 1990 period would do so in 1975 (1977) if the coal ban is enacted. Based on the above analysis this assumption represents an overestimation of the abandoned buildings, since some of the buildings would not have been abandoned for many years and buildings that would not be abandoned without the policy would have no additional incentives to abandon with the policy.

[a]A higher vacancy rate was observed in buildings still using coal than in buildings that were converted from coal (see Appendix B).

NOTES TO APPENDIX D

1. U.S. Bureau of the Census, *1970 Census of Housing*, Final Report HC(2)–44, *Metropolitan Housing Characteristics, Chicago, Illinois, SMSA* (Washington: U.S. Government Printing Office, 1972), p 54.
2. Institute of Real Estate Management, *Apartment Building Income–Expense Analysis 1970* (Chicago: National Association of Real Estate Boards, 1970).
3. U.S. Bureau of the Census, *1970 Census of Housing*, Vol. 1, Part 15 (Illinois) (Washington: U.S. Government Printing Office, 1972), p 195.

Appendix E

Estimated Elasticities of Fuel Use with Respect to Fuel Price

The purpose of this appendix is to estimate the sensitivity of conversions and fuel usage to changes in fuel prices and, correspondingly, in space heating costs. Recent trends in coal prices and potential changes of gas and oil prices are expected to modify the observed trends and tendencies in fuel usage for space heating. In this appendix we attempt to quantify the magnitudes of these influences. The quantification is based on changes that took place in the past. The sensitivity of fuel used for space heating to fuel prices is estimated using 3 different models. The first, estimating demand elasticities for space heating fuels, uses United States data. The other two, estimating demand elasticities for conversion from coal, use Chicago data.

The first model utilizes a modified demand equation for space heating fuel. The three unrelated equations that were estimated contain the following explanatory variables: (1) the prices of the fuels,[1] which are the factors determining annual fuel costs; (2) per capita income, which is a proxy for labor costs; (3) January average temperature; (4) percentage of single-family homes; and (5) the percentage of dwelling units that were built after 1950. The dependent variables were the percentages of dwelling units heated by gas, oil, and coal, respectively.

The results of the log-linear regression equations are presented in Table E-1. The estimated coefficients are the corresponding elasticities. The sample consists of 40 observations (one from 20 United States cities in 1960 and 1970).[2]

In the following, we compare the outcome with the expected results. The comparison is related only to the sign of the elasticities since no prior expectation can be formulated with respect to their magnitude. The signs of the price elasticities are as expected—negative and mostly significant (1 percent). Cross price elasticities are expected to be positive (substitutes). We found some to be negative, but they are insignificant ($t < 1.0$). The income

Table E-1. Regression Coefficients for Space Heating Fuel Equations

	Dependent Variables					
	1 *Percent Gas*		*2* *Percent Oil*		*3* *Percent Coal*	
Explanatory Variables						
Price of gas	-1.600 (.250)	-1.621 (.217)	3.878 (.486)	3.732 (.505)	2.134 (.983)	1.129 (.718)
Price of oil	-.618 (1.82)	-.004 (1.099)	-10.095 (3.518)	-10.909 (3.936)	-11.226 (7.116)	-5.046 (5.597)
Price of coal	-.658 (.575)	-.245 (.493)	.868 (1.114)	1.123 (1.149)	-3.226 (2.254)	-2.214 (1.633)
Income		.689 (.378)		.422 (.880)		-2.015 (1.251)
January Temperature		-.462 (.320)		.384 (.746)		-1.867 (1.061)
% Single Family Homes		-.450 (.179)		.113 (.419)		-.658 (.596)
% New Dwelling Units		.323 (.198)		-.806 (.462)		-1.448 (.656)
Corrected R^2	.542	.689	.677	.684	.334	.581

Note: Values in parentheses are standard errors. The form of the equations is log-linear.

elasticity is expected to be positive for gas and oil and negative for coal for the following reasons:

1. Higher incomes imply higher labor costs that mainly affect coal usage.
2. The higher the income, the greater the preference for clean fuels.
3. The higher the income, the higher are rents, thus, cash or financial shortages for investment in conversions are less severe.
4. The higher the income, the higher is the growth rate—i.e., there are relatively more new homes. This effect is also partly captured by the variable "percentage of new dwelling units."

The result for the sign of the income variable confirms this expectation.

For January temperature we did not have strong a priori expectations, although for gas we expected a positive coefficient and for coal a negative one. The reason is that in relatively warm climates the cost of maintaining indoor temperature at a comfortable and stable level is higher with coal than with gas. The gas result does not confirm our anticipation.

The coefficient of the single-family home variable is expected to be positive for gas and oil and negative for coal, since furnace efficiencies for gas and oil do not vary greatly with building size while for coal the efficiency increases significantly with building size. The result in the gas equation does not confirm this expectation.

A large proportion of new dwelling units is expected to have a positive effect on the rate of gas usage and a negative one for oil and coal. The reason is a historical one that relates to the availability of gas in large magnitudes. The results support this hypothesis.

In general, the results support the basic hypothesis that fuel use is sensitive to space heating costs. The elasticities estimated are long-run elasticities—i.e., a cross-section estimation. These elasticities are thus, a priori, expected to be higher than those measured in the city of Chicago in the 1968 to 1971 period, since the latter are short-run elasticities. Note that this model does not contain conversion costs as a factor in fuel use. We could have included it by estimating a relationship between the differences in fuel use in 1960 and 1970, if we had had good reason to believe that conversion costs differ from one city to the other.

The second model estimated from Chicago data includes as explanatory variables the annual space heating cost differential between coal and gas, which is expected to have a positive coefficient, and conversion costs from coal to gas, expected to have a negative coefficient. The dependent variable is the rate of decline of coal-using dwelling units by building size in the 1962 to 1968 and 1968 to 1971 periods. A log-linear relation was estimated;

$$RDC = e^{6.958} (H_c - H_g)^{.634} C^{-1.168}$$

$$(.188) \quad (.257)$$

corrected $R^2 = .502$

where

$RDC \equiv$ Rate of decline of coal using dwelling units;

$H_c \equiv$ Heating costs using coal;

$H_g \equiv$ Heating costs using gas;

$C \equiv$ Conversion costs from coal to gas.

Altogether, there were 22 observations (single-family homes were excluded). When building size dummy variables were included the cost differential elasticity did not change [.661(.360)], while the conversion costs coefficient became positive but insignificant [.275(3.09)], a result that one could hardly accept. If all the relevant variables were included in the rate of conversion model, one would expect the elasticity of the heating cost differential to be relatively high. The estimated elasticity, however, was less than unity. These two deficiencies led us to reject these results for the policy costs evaluations.

The third model calculates an arc elasticity of a conversion rate with respect to the ratio of the annual space heating cost differential between coal and gas to conversion costs. Table E–2 contains the relevant calculations and results. The elasticities calculated are based on 1971 prices. The elasticity is

$$\eta = \frac{\Delta \text{ Conversion Rate}}{\text{Average Conversion Rate}} \Bigg/ \frac{\Delta \text{ (Heating Cost Differential/ Conversion Costs)}}{\text{Average (Heating Cost Differential / Conversion Costs)}}$$

The data are by building size for group central heating. Gas shortages probably lowered the estimated elasticity. If one relates the observed conversions only to fuel costs (disregarding labor cost, which some landlords view as a fixed cost), the estimated elasticities of conversion are much lower. A methodological problem that comes up is how to explain conversions in the 1962 to 1966 period when the fuel costs of coal were less than gas. The method of calculation actually assigns an equal elasticity to the heating cost differential and the conversion cost.

The results show that the elasticities for all building sizes do not deviate strongly from unity. The method used to calculate the effect of changes in fuel prices on conversion rates employed unitary elasticities by assuming

Table E-2. Elasticities of Residential Conversions from Coal to Gas in Chicago, 1962 to 1972

Number of Dwelling Units	*Coal–Gas Heating Cost Differential, (1966 Prices)*	*Heating Cost Differential/ Conversion Cost (1966 Prices)*	*Coal–Gas Heating Cost Differential (1971 Prices)*	*Heating Cost Differential/ Conversion Cost (1971 Prices)*	*Arc Elasticity 1969–1972*
2	$ 201.4	.2718	$ 483.3	.5686	.322
3	213.6	.2119	649.5	.5668	.337
4	265.6	.2231	765.8	.5631	.663
6	359.1	.2418	999.9	.5611	1.876
10	538.1	.2374	1462.4	.5735	1.238
16	780.9	.2581	2139.3	.6307	1.342
25	1117.8	.3299	3169.2	.8353	1.490
34	1446.7	.3743	4296.8	.9951	1.296
47	1620.5	.3471	5136.3	.9845	1.297
100	1452.5	.1759	7090.8	.7707	.676

proportionality between the fuel price differential and the change in the number of dwelling units that use coal under the different prices (see Appendix G).

NOTES TO APPENDIX E

1. The fuel prices used for 1960 were the January, 1959 prices and for 1970 the January, 1964 prices (1964 was the last year coal prices were reported separately). See U.S. Bureau of Labor Statistics, *Retail Prices and Indices of Fuels and Electricity* (Washington: U.S. Government Printing Office, Monthly).
2. U.S. Bureau of the Census, *1970 Census of Housing*, Vol. 1, Part 1, *United States Summary* (Washington: U.S. Government Printing Office), Table 41.

Appendix F

Detailed Components of Estimated Policy Costs with Constant Fuel Prices

This appendix contains the results of the calculations of the social costs related to various fuel policies. For the sulfur law and the 1975 coal ban each cost component is represented by a table indicating the annual costs up to the year 1990. Values for each policy and for a less restrictive policy are presented in each table, and the difference in costs—i.e., the marginal costs—are given. For the other policies summary marginal cost tables are presented.

COSTS RELATED TO THE LOW SULFUR LAW

The low sulfur law passed by the city of Chicago in 1968 came into effect simultaneously with major changes in the coal market. Thus, the observed change in the price per ton of coal in Chicago from $15.45/ton in October 1968 to $33.35/ton in January 1972 cannot be attributed entirely to the low sulfur law. One has to take into account the different types of coal used in the two time periods—i.e., eastern Kentucky and southern Illinois coal. These types of coal have different heat contents, namely 26 million Btu per ton, compared with 23 million Btu per ton, respectively. Furthermore, one has to consider the general upward trend in coal prices in the United States in the 1969 to 1971 period.

Table F-1 contains the essential information on the trend in coal prices. These are the prices of commercial stoker nut defined for an 8-ton load with driver and hiker costs added. These few numbers tell us an important story. The price of southern Illinois coal increased by about 50 percent in a period of a declining demand for that coal. The price of eastern Kentucky coal increased by 60 percent at a time of increasing demand. The first conclusion is straightforward. The increased conversion from coal in the 1969 to 1972 period is only partly the result of the low sulfur law. The major factor affecting this process is the general increase in the price of coal. Nationally, one finds an

Table F-1. Coal Prices in Chicago (Dollars per Ton)

	Heating Season		
	1969	*1970*	*1971*
Eastern Kentucky	20.70	32.15	33.35
Southern Illinois	16.75	25.00	25.00

Source: Quoted by Chicago coal dealer.

increase of 62 percent in the wholesale price of bituminous coal between 1969 and 1971. Obviously the low sulfur law in Chicago has hardly affected this price change.

The above discussion is significant for the evaluation of the costs and benefits of the low sulfur law. The relevant question relates to the background situation to which the effects of the law have to be compared. The relevant background is an unobserved trend that is somewhere between the extrapolated 1962 to 1968 trend and the observed 1968 to 1972 trend. Hence, it was necessary to subtract the effect of the low sulfur law from the observed 1968 to 1972 trend in order to get the relevant background trend.

The method used for estimating the hypothesized 1968 to 1972 trend without a sulfur law is a linear interpolation between the two trends. An alternative method for determining the influence of fuel price changes on conversions from coal would be to use the corresponding elasticities. Some of the difficulties associated with the latter approach are discussed in Appendix E. To use the linear interpolation approach the ratio $P = AB/AC$ was defined as (see Figure F-1)

$$P = \frac{P_1 - P_0}{P_2 - P_0} = \frac{30.58 - 20.66}{33.35 - 20.66} = .782,$$

where

$P_0 \equiv$ the estimated 1971 price of an equivalent ton[a] of Illinois coal if its price increased in the 1969 to 1971 period at the same rate it increased in the 1964 to 1968 period $[P_0 = (18.28)(26/23) = 20.66]$;[b]

[a]An equivalent ton of coal is defined as the quantity of coal having a heat value of 26×10^6 Btu's. Illinois coal has an average heat content of 23×10^6 Btu's per ton. Therefore, the price of an equivalent ton of Illinois coal is (26/23) times the price of a ton of Illinois coal.

[b]Coal prices are from Table 2-4.

P_1 ≡ the estimated 1971 price of an equivalent ton of high sulfur Illinois coal if its price increased in the 1969 to 1971 period at the same rate as coal prices increased nationally in the 1969 to 1971 period [$P_1 = (27.05)(26/23) = 30.58$];

P_2 ≡ the observed 1971 price of a ton of low sulfur eastern Kentucky coal ($P_2 = 33.35$). (This coal has an average heat content of 26×10^6 Btu's per ton.)

One can see that the low sulfur law generated an increase in coal price of only $2.77 per 26 million Btu's—i.e., $P_2 - P_1$. The ratio 0.782 implies that 78 percent of the difference between the two trends of coal using dwelling units would have occurred without the sulfur law. Hence, the costs charged to the sulfur law and the benefits attributed to it are considerably smaller than those calculated using the extrapolated 1962 to 1968 trend as a reference point.

Two additional remarks are appropriate before presenting the cost data: 1. The fuel prices for the 1969 to 1971 period were the *actual* fuel prices for oil and gas, and the calculated price for an equivalent ton of Illinois coal. Fuel prices for the 1972 to 1990 period are assumed to be the 1971 prices. 2. Extra abandonment effects were not attributed to the low sulfur law, as was done for the fuel ban policies.

Four cost tables are presented for the sulfur law. Values for the sulfur law and the no control policies are presented and the difference in their

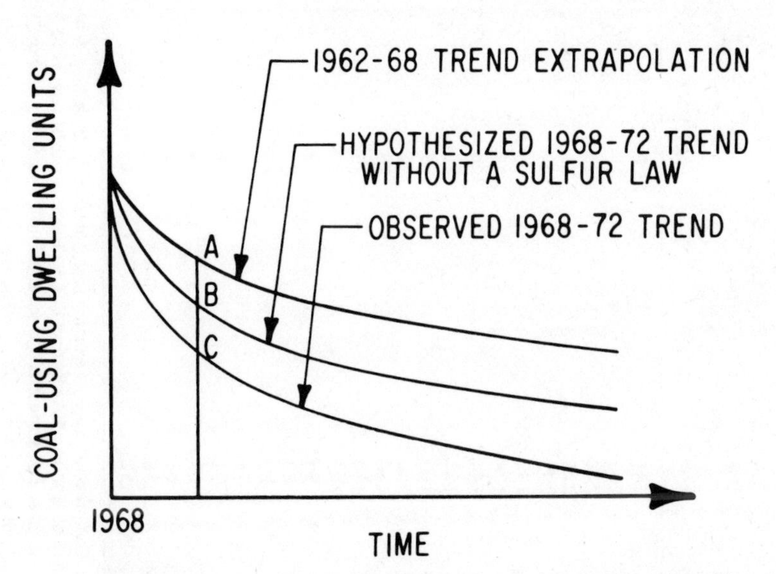

Figure F-1. Three Trends of Coal Use in the 1968 to 1972 Period.

Table F-2. Coal-Using Dwelling Units and Coal Sales with and without the Sulfur Law (SL), 1968 to 1990 (Thousands)

Year	Dwelling Units 1962–68 Trend	Dwelling Units With SL	Difference	Difference X.218	Tons Coal Used 1962–68 Trend	Tons Coal Used With SL	Difference	Difference X.218
1968	281	281[a]	0	0.0	1684[b]	1684[b]	0	0.0
1969	262	254[a]	8	1.7	1541[b]	1406[b]	135	29.4
1970	241	215[a]	26	5.7	1410[b]	1174[b]	236	51.4
1971	223	187[a]	36	7.8	1142[c]	868[c]	274	59.7
1972	207	138[a]	69	15.0	1045	725	320	69.8
1973	192	117	75	16.4	929	571	358	78.0
1974	178	100	78	17.0	858	485	373	81.3
1975	164	84	80	17.4	791	411	380	82.8
1976	152	72	80	17.4	728	348	380	82.8
1977	140	61	79	17.2	671	295	376	82.0
1978	130	51	79	17.2	619	250	369	80.4
1979	120	43	77	16.8	569	210	359	78.3
1980	110	37	73	15.9	525	177	348	75.9
1981	103	31	72	15.7	471	143	328	71.5
1982	95	26	69	15.0	431	119	312	68.0
1983	89	22	67	14.6	395	100	295	64.3
1984	82	18	64	14.0	361	83	278	60.6
1985	76	15	61	13.3	330	70	260	56.7
1986	71	13	58	12.6	302	58	244	53.2
1987	65	11	54	11.8	277	49	228	49.7
1988	61	9	52	11.3	253	41	212	46.2
1989	56	8	48	10.5	231	34	197	42.9
1990	52	6	46	10.0	212	28	184	40.1

Note: The present value of earning losses in coal distribution, which equals the extra cost to be borne by society for additional capital and manpower in oil and gas distribution due to the low sulfur law, is $1.93 million for a 10 percent discount rate and $1.01 million for a 20 percent discount rate. The annualized costs are $.197 and $.171 million, respectively. The assumed loss per ton of coal is $3 per ton.

[a] Actual numbers (Peoples Gas, Light and Coke Company).

[b] Illinois coal.

[c] Eastern Kentucky coal. This relates to all years after 1971, although given no low sulfur law, Illinois coal would have been 1.13 times the number given in the table.

costs–i.e., the marginal costs–are given. Table F-2 contains the parameters needed to calculate the loss of returns to human and non-human capital presently occupied in the distribution of coal. These are the number of dwelling units using coal and the quantity of coal sold annually. The 1971 mark-up on eastern Kentucky coal was $6.73 per ton. Part of this mark-up is payments for variable costs (trucks, gasoline, repairs, and so forth) that are saved if no coal is distributed. The returns to human and non-human capital occupied in coal distribution that are non-mobile or non-adaptable to other uses were assumed to be $3.00 per ton, which is about 10 percent of the sale price. Table F-3 contains the annual social costs due to conversions. Table F-4 contains the annual social costs related to changes in the fuel used. In Chicago, these costs are negative under the assumption of constant fuel prices–i.e., they are savings. Table F-5 summarizes the costs (savings) components of the fuel policy provided in the previous tables. Each of the components is presented in terms of a present value and an annualized value. The summary table is divided into various sections that reflect the parametric analyses conducted for the sulfur law.

Table F-3. Costs of Conversion from Coal with and without the Sulfur Law (SL), 1969 to 1990 (Millions of Dollars)

Year	*Without SL (1962-68 Trend)*	*With SL*	*Net Cost (Savings)*[a]
1969	4.41	8.17	3.76
1970	4.07	6.92	2.85
1971	3.76	5.86	2.10
1972	3.47	4.96	1.49
1973	3.15	4.24	1.09
1974	2.92	3.52	.60
1975	2.76	3.02	.26
1976	2.55	2.53	(.02)
1977	2.35	2.16	(.19)
1978	2.09	1.84	(.25)
1979	2.01	1.55	(.46)
1980	1.78	1.31	(.47)
1981	1.68	1.11	(.57)
1982	1.55	.94	(.61)
1983	1.43	.80	(.63)
1984	1.32	.68	(.64)
1985	1.21	.57	(.64)
1986	1.12	.48	(.64)
1987	1.04	.41	(.63)
1988	0.95	.35	(.60)
1989	0.88	.29	(.59)
1990	0.81	.25	(.56)

[a]Only .218 of the difference is due to the low sulfur law. The present value of the low sulfur law's share is $1.70 million for a 10 percent discount rate and $1.58 million for a 20 percent discount rate. The annualized costs are $.17 and $.27 million, respectively.

Table F-4. Fuel Costs with and without the Sulfur Law, 1969 to 1990 (Millions of Dollars)

Year	α	β	λ	*Net Costs*[a] *(Savings)*
1969	3.44	2.69	.76	.01
1970	12.08	7.13	2.18	(2.77)
1971	14.34	9.24	3.02	(2.08)
1972	10.53	6.79	2.81	(.93)
1973	8.93	5.76	2.60	(.57)
1974	7.59	4.89	2.41	(.29)
1975	6.43	4.15	2.22	(.06)
1976	5.46	3.52	2.05	.11
1977	4.63	2.99	1.39	.25
1978	3.91	2.52	1.74	.35
1979	3.30	2.21	1.60	.51
1980	2.36	1.80	1.48	.92
1981	2.00	1.55	1.37	.92
1982	1.69	1.32	1.26	.89
1983	1.43	1.12	1.16	.85
1984	1.21	.95	1.07	.81
1985	1.03	.80	.99	.76
1986	.87	.68	.91	.72
1987	.74	.58	.84	.68
1988	.61	.49	.78	.66
1989	.52	.41	.73	.62
1990	.44	.34	.69	.59

Note: See the last section of this appendix for the mathematical derivation of the formula used to calculate net costs–i.e., Δ cost = $\alpha - \beta - \lambda$. It is assumed for purposes of this table that $P_t = .782$ for all t in Equation 15. The data are based on projected coal prices if the sulfur law was not enacted.

[a] The present value of the fuel costs of the low sulfur law is negative–i.e., fuel savings are generated. The present value of the fuel savings is $2.73 million for a discount rate of 10 percent and $3.23 million for 20 percent. The corresponding annualized savings are $.28 and $.55 million, respectively.

COSTS RELATED TO THE 1975 COAL BAN

Four cost tables are presented for the coal ban. The base case is the sulfur law. Tables F-6 through F-8 provide the data for the annual losses in coal distribution, conversion costs, and fuel costs (savings), respectively. Table F-9 summarizes the costs. The summary table presents the costs for the parametric analyses conducted for the 1975 coal ban.

COSTS RELATED TO OTHER FUEL POLICIES

Only the summary tables for the 1977 coal ban, the 1975 coal and oil ban and the 1977 coal and oil ban are presented. These are Tables F-10 through F-12, respectively.

Table F-5. Summary of Costs of Sulfur Law Relative to the No Control Policy (Millions of Dollars)

	Present Value 10%	*Annualized 10%*	*Present Value 20%*	*Annualized 20%*
I. Basic Assumptions.				
Coal Marketing Losses @ $3/Ton	1.93	.20	1.01	.17
Cost of Earlier Furnace Conversions	1.70	.17	1.58	.27
Additional Fuel Costs	(2.73)	(.28)	(3.23)	(.55)
Total Costs	.90	.09	(.64)	(.11)
II. As I, but costs of conversions are twice those presented in Table 2-2.				
Coal Marketing Losses	1.93	.20	1.01	.17
Cost of Earlier Furnace Conversions	3.41	.35	3.17	.27
Additional Fuel Costs	(2.73)	(.28)	(3.23)	(.55)
Total Costs	2.61	.27	.95	.15

Note: The present value is calculated for 1973. Figures in parentheses are negative costs or savings.

Table F-6. Coal-Using Dwelling Units and Coal Sales with and without a 1975 Coal Ban (CB), 1973 to 1990 (Thousands)

	Dwelling Units			*Tons of Coal Sold*		
Year	*Without CB*	*With CB*	*Difference*	*Without CB*	*With CB*	*Difference*
1973	117.0	90.0	27.0	571	440	131
1974	100.0	44.0	56.0	485	216	269
1975	84.0	0.0	84.0	411	0	411
1976	72.0	0.0	72.0	348	0	348
1977	61.0	0.0	61.0	295	0	295
1978	51.0	0.0	51.0	250	0	250
1979	43.0	0.0	43.0	210	0	210
1980	37.0	0.0	37.0	177	0	177
1981	30.6	0.0	30.6	143	0	143
1982	25.8	0.0	25.8	119	0	119
1983	21.8	0.0	21.8	100	0	100
1984	18.4	0.0	18.4	83	0	83
1985	15.5	0.0	15.5	70	0	70
1986	13.1	0.0	13.1	58	0	58
1987	11.0	0.0	11.0	49	0	49
1988	9.3	0.0	9.3	41	0	41
1989	7.8	0.0	7.8	34	0	34
1990	6.0	0.0	6.0	28	0	28

Note: Multiplying the difference in the tons of coal sold by $3.00 provides the annual value of these lost resources. The present value of this cost is $5.48 million for a discount rate of 10 percent and $3.92 million for a rate of 20 percent. The corresponding annualized costs are $.61 and $.68 million.

Table F-7. Costs of Conversion from Coal with and without a 1975 Coal Ban (CB), 1973 to 1990 (Millions of Dollars)

Year	Without CB (SL)	With CB	Net Cost (Savings)[a]
1973	4.24	9.14	4.90
1974	3.52	8.97	5.45
1975	3.02	9.00	5.98
1976	2.53	0.0	(2.53)
1977	2.16	0.0	(2.16)
1978	1.84	0.0	(1.84)
1979	1.55	0.0	(1.55)
1980	1.31	0.0	(1.31)
1981	1.11	0.0	(1.11)
1982	.94	0.0	(.94)
1983	.80	0.0	(.80)
1984	.68	0.0	(.68)
1985	.57	0.0	(.57)
1986	.48	0.0	(.48)
1987	.41	0.0	(.41)
1988	.35	0.0	(.35)
1989	.29	0.0	(.29)
1990	.25	0.0	(.25)

Note: The assumptions underlying the conversions from coal are:

1. Oil conversions to gas with and without a coal ban are the same.
2. Conversions from coal are distributed 89 percent to gas and 11 percent to oil. These two figures are the relative proportions predicted for gas and oil in 1980 given the 1968 to 1971 trends.

[a]The marginal present value of the 1975 coal ban with a discount rate of 10 percent is \$6.64 million, and for a 20 percent discount rate, \$8.61 million. The corresponding annualized costs are \$.74 and \$1.49 million, respectively.

WORKING EQUATIONS FOR FUEL SAVINGS UNDER THE LOW SULFUR LAW

Notation

N ≡ Number of coal users in 1968

C_t ≡ Number of coal users at time t without SL—i.e., pre-1968 trend

C_t^* ≡ Number of coal users at time t with SL

PC_t ≡ Price of coal at time t without SL

PC_t^* ≡ Price of coal at time t with SL

PG_t ≡ Price of "other" fuels at time t (unaffected by policy)

F_t ≡ Fuel cost at time t without SL

F_t^* ≡ Fuel cost at time t with SL

Table F-8. Reductions in Fuel Costs with and without a 1975 Coal Ban (CB), 1973 to 1990 (Millions of Dollars)

Year	*Without CB*	*With CB*	*Net Costs (Savings)*
1973	1.43	3.29	(1.86)
1974	2.66	6.43	(3.77)
1975	3.70	9.47	(5.77)
1976	4.58	9.47	(4.89)
1977	5.33	9.47	(4.14)
1978	5.97	9.47	(3.50)
1979	6.52	9.47	(2.95)
1980	6.99	9.47	(2.48)
1981	7.35	9.47	(2.12)
1982	7.65	9.47	(1.82)
1983	7.94	9.47	(1.53)
1984	8.17	9.47	(1.30)
1985	8.39	9.47	(1.08)
1986	8.59	9.47	(.88)
1987	8.76	9.47	(.71)
1988	8.90	9.47	(.57)
1989	9.04	9.47	(.43)
1990	9.16	9.47	(.31)

Note: Assume oil conversions with and without coal ban are the same. The present value of this savings is $24.67 million for a discount rate of 10 percent and $16.55 million for a rate of 20 percent. The annualized savings are $2.74 and $2.86 million, respectively.

Fuel Savings without Coal Price Change

Define

$$C_o \equiv N, \tag{1}$$

$$\Delta C_t^{t-1} \equiv C_{t-1} - C_t, \text{ and} \tag{2}$$

$$\Delta C_t^o \equiv \sum_{j=1}^{t} \Delta C_j^{j-1} \tag{3}$$

Note that

$$C_t + \Delta C_t^o = C_t^* + \Delta C_t^{*o} = N, \text{ for all } t, \tag{4}$$

thus

$$C_t = C_t^* + \Delta C_t^{*o} - \Delta C_t^o. \tag{4'}$$

Table F-9. Summary of Costs of Coal Ban in 1975 Relative to the Sulfur Law (Millions of Dollars)

	Present Value 10%	*Annualized 10%*	*Present Value 20%*	*Annualized 20%*
I. Basic Assumptions.				
Coal Marketing Losses @ $3/Ton	5.48	.61	3.92	.68
Cost of Earlier Furnace Conversions	6.64	.74	8.61	1.49
Additional Fuel Costs	(24.67)	(2.74)	(16.55)	(2.86)
Total Costs	(12.55)	(1.39)	(4.02)	(.69)
II. As I, but costs of conversions are twice those presented in Table 2-2.				
Coal Marketing Losses	5.48	.61	3.92	.68
Cost of Earlier Furnace Conversions	13.28	1.47	17.22	2.98
Additional Fuel Costs	(24.67)	(2.74)	(16.55)	(2.86)
Total Costs	(5.91)	(.66)	4.60	.80
III. As II, but with premature abandonments.				
Coal Marketing Losses	5.22	.58	3.67	.63
Cost of Earlier Furnace Conversions	9.46	1.06	12.10	2.10
Moving, Rent, and Inconveniences	26.55	2.95	16.94	2.93
Additional Fuel Costs	(22.96)	(2.55)	(14.93)	(2.58)
Total Costs	18.27	2.03	17.78	3.08

Note: Present values are for 1973. Figures in parentheses are negative costs—i.e., savings.

Furthermore,

$$F_t = (PC_t)\, C_t + (PG_t)\, \Delta C_t^o, \tag{5}$$

$$F_t^* = (PC_t^*)\, C_t^* + (PG_t)\, \Delta C_t^{*o}, \text{ and} \tag{6}$$

$$\begin{aligned} F_t^* - F_t &= (PC_t^*)\, C_t^* + (PG_t)\, \Delta C_t^{*o} - (PC_t)\, C_t - (PG_t)\, \Delta C_t^o \\ &= (PC_t^*)\, C_t^* + (PG_t)\, [\Delta C_t^{*o} - \Delta C_t^o] - (PC_t)\, C_t. \end{aligned} \tag{7}$$

Substitute (4′) into (7)

$$\begin{aligned} F_t^* - F_t &= (PC_t^*) C_t^* + (PG_t)\, [\Delta C_t^{*o} - \Delta C_t^o] - (PC_t)\, [C_t^* + \Delta C_t^{*o} - \Delta C_t^o] \\ &= (PC_t^*)\, C_t^* + (PG_t)\, [\Delta C_t^{*o} - \Delta C_t^o] - (PC_t)\, C_t^* - (PC_t)\, [\Delta C_t^{*o} - \Delta C_t^o] \\ &= [PC_t^* - PC_t]\, C_t^* + [PG_t - PC_t]\, [\Delta C_t^{*o} - \Delta C_t^o]. \end{aligned} \tag{8}$$

Table F-10. Summary of Costs of Coal Ban in 1977 Relative to the Sulfur Law (Millions of Dollars)

	Present Value 10%	*Annualized 10%*	*Present Value 20%*	*Annualized 20%*
I. Basic Assumptions.				
Coal Marketing Losses	3.64	.40	2.39	.41
Costs of Earlier Furnace Conversions	4.51	.50	5.20	.90
Additional Fuel Costs	(17.17)	(1.91)	(11.17)	(1.93)
Total Costs	(9.02)	(1.01)	(3.58)	(.62)
II. As I, but costs of conversions are twice those presented in Table 2-2.				
Coal Marketing Losses	3.64	.40	2.39	.41
Cost of Earlier Furnace Conversions	9.02	1.00	10.40	1.80
Additional Fuel Costs	(17.17)	(1.91)	(11.17)	(1.93)
Total Costs	(4.51)	(.51)	1.62	.28
III. As II, but with premature abandonments.				
Coal Marketing Losses	3.15	.35	1.95	.34
Cost of Earlier Furnace Conversions	5.42	.60	6.16	1.06
Moving, Rent, and Inconveniences	12.12	1.35	7.86	1.36
Additional Fuel Costs	(14.36)	(1.59)	(8.78)	(1.52)
Total Costs	6.33	.71	7.19	1.24

Note: Present values are for 1973. Figures in parentheses are negative costs–i.e., savings.

In Equation 8 we have

$PC_t^* - PC_t \equiv$ the increase in the cost of coal

$PG_t - PC_t \equiv$ the fuel cost difference to coal users converting to other fuels.

From Equation 4 we have

$$\Delta C_t^{*o} - \Delta C_t^o = C_t - C_t^* . \tag{9}$$

Substitute (9) into (8)

$$F_t^* - F_t = [PG_t - PC_t]\ C_t + [PC_t^* - PG_t]\ C_t^* . \tag{10}$$

From Tables 4-2 and 4-5 we see that for gas conversions $(PG_t - PC_t) < 0$ and $(PC_t^* - PG_t) > 0$ for $t > 1969$. Since $PC_t^* > PC_t$, we know that $|PG_t - PC_t| \leqslant |PC_t^* - PG_t|$. Furthermore, $C_t > C_t^*$ for all t. Therefore, a change in the sign of $F_t^* - F_t$ is possible over time.

Table F-11. Summary of Costs of Coal and Oil Ban in 1975 Relative to the Sulfur Law (Millions of Dollars)

	Present Value 10%	*Annualized 10%*	*Present Value 20%*	*Annualized 20%*
I. Basic Assumptions.				
Coal Marketing Losses @ $3/Ton	5.48	.61	3.92	.68
Oil Marketing Losses @ $.01/Gal.	7.49	.83	4.98	.86
Cost of Earlier Furnace Conversions from Coal	6.64	.74	8.61	1.49
Cost of Earlier Furnace Conversions from Oil	33.10	3.67	34.44	5.96
Additional Fuel Costs	(61.45)	(6.82)	(42.39)	(7.33)
Total Costs	(8.74)	(.97)	9.56	1.65
II. As I, but costs of conversions are twice those presented in Table 2-2.				
Coal Marketing Losses	5.48	.61	3.92	.68
Oil Marketing Losses	7.49	.83	4.98	.86
Cost of Earlier Furnace Conversions from Coal	13.28	1.48	17.22	2.98
Cost of Earlier Furnace Conversions from Oil	66.20	7.34	68.88	11.92
Additional Fuel Costs	(61.45)	(6.82)	(42.39)	7.33
Total Costs	31.00	3.44	52.61	9.10

Note: Present values are for 1973. Figures in parentheses are negative costs–i.e., savings.

Table F-12. Summary of Costs of Coal and Oil Ban in 1977 Relative to the Sulfur Law (Millions of Dollars)

	Present Value 10%	*Annualized 10%*	*Present Value 20%*	*Annualized 20%*
I. Basic Assumptions.				
Coal Marketing Losses $3/Ton	3.64	.40	2.39	.41
Oil Marketing Losses $.01/Gal.	6.10	.68	3.76	6.50
Cost of Earlier Furnace Conversions from Coal	4.51	.50	5.20	.90
Cost of Earlier Furnace Conversions from Oil	26.05	2.89	26.82	4.63
Additional Fuel Costs	(46.90)	(5.21)	(29.67)	(5.13)
Total Costs	(6.60)	(.73)	8.50	1.47
II. As I, but costs of conversions are twice those presented in Table 2-2. The corresponding total costs are:				
	23.96	2.66	40.52	7.01

Note: Present values are for 1973. Figures in parentheses are negative costs–i.e., savings.

Fuel Savings with an Exogenous Coal Price Change

Let the prime denote variables in the 1968-90 period without a sulfur law, but with exogenous coal price increases. Then from Equation 10

$$\Delta \text{Cost} = F_t^* - F_t' = [PG_t - PC_t'] \, C_t' + [PC_t^* - PG_t] \, C_t^*. \tag{11}$$

Define

$$C_t' \equiv C_t - P_t \, (C_t - C_t^*) \text{ and} \tag{12}$$

$$P_t = \frac{PC_t' - PC_t}{PC_t^* - PC_t}. \tag{13}$$

Substitute (12) into (11)

$$\Delta \text{ Cost} = [PG_t - PC_t'] \, C_t - P_t \, [PG_t - PC_t'] \, (C_t - C_t^*) + [PC_t^* - PG_t] \, C_t^*$$

$$= [PG_t - PC_t'] \, C_t - P_t \, [PG_t - PC_t'] \, C_t + P_t [PG_t - PC_t'] \, C_t^* + [PC_t^* - PG_t] \, C_t^*$$

$$= (1 - P_t) \, [PG_t - PC_t'] \, C_t + P_t \, [PG_t - PC_t'] \, C_t^* + [PC_t^* - PG_t] \, C_t^* \tag{14}$$

Rearranging terms, Equation 14 becomes

$$\Delta \text{ Cost} = [PC_t^* - PG_t] \, C_t^* - P_t \, [PC_t' - PG_t] \, C_t^* - (1 - P_t) \, [PC_t' - PG_t] \, C_t \tag{15}$$

$$= \alpha - \beta - \lambda.$$

Appendix G

Policy Costs with Changing Fuel Prices

To determine the social costs of a fuel policy under non-constant fuel prices (primarily gas) two types of price changes must be considered. The first occurs if a rightward shift of the local demand for gas, which is caused by a fuel policy, results in an intersection between the local demand for gas and the local supply of gas on the upward sloping part of the supply curve. The price of gas increases for all consumers, including consumers that were using gas prior to the enactment of the policy. This will be called an endogenous price change. The second is a gas price increase that is independent of the fuel policy. An example would be price increases resulting from relaxing controls on the well-head price of gas. This will be called an exogenous price change.

ENDOGENOUS PRICE CHANGES

The appropriate supply curve of gas to evaluate endogenous price changes is the adjusted long-run winter supply curve the consumers in Chicago are facing. (Recall that the gas utility is regulated, thus restricting the rate of profit it is allowed to make. For simplicity, assume that the supply function the consumers are facing is already adjusted for this constraint on returns to capital.) The shape of this curve depends on the shape of the supply curve the gas utility is facing, and the winter gas distribution function. The upward sloping part of the supply curve faced by the gas consumers (see Figure 2-4) reflects increased marginal costs of storing gas in the summer for winter consumption. These increasing costs result from constraints on storage space requiring the use of marginal storage space and liquifying and regasifying summer gas. The demand for gas shifts continuously upward as the number of gas customers increases. As demand shifts from D_0 to D_3 the price of gas increases from P_0 to P_3.

In the cost calculations for residential fuel policies presented in Appendix F, constant 1971 heating season prices of gas—i.e., for January 1972–

were assumed to prevail in the future. In the following the additional social costs imposed on the residential sector in the city of Chicago due to an upward sloping supply curve of gas are discussed. It should be noted that endogenous gas price increases also affect non-residential gas users. Since total gas consumption by the residential sector in the fourth quarter of 1971 and the first quarter of 1972 (winter 1971) was 50.3 percent of the winter gas consumption in Chicago, a rough estimation that includes all the other sectors renders a cost that is about twice that for the residential sector.[a]

Since the number of conversions to gas under the low sulfur law would eventually equal the number that converted under a fuel ban, the notion of an upward sloping long-run winter gas supply curve implies that the coal and coal and oil bans would cause a premature increase in the price of gas in Chicago. Because these increases occur earlier in time the present value and the annualized cost of the policy increase. In this sense, the effect of an endogenous price increase on the social cost of the policy is identical to the effects of earlier conversion or abandonments.

The median size coal- and oil-using buildings in Chicago in 1972 had 22 and 2 dwelling units, respectively. The average consumption of natural gas per dwelling unit in these size buildings is $.833 \times 10^3$ and 1.499×10^3 therms, respectively. (This was calculated by using the engineering cost model reported in Appendix C.) By the summer of 1972 there were 138 thousand coal-using dwelling units and 162 thousand oil-using units in Chicago.

The social cost associated with endogenous gas price increases is calculated assuming: (1) there are no abandonments; (2) 89 percent of the coal-using dwelling units existing in 1972 convert to gas and 11 percent to oil under a coal ban and all of the coal- and oil-using dwelling units existing in 1972 convert to gas under the coal and oil ban; and (3) the quantity of gas demanded in the winter by the other sectors in Chicago remains constant at their 1971 winter consumption level. Finally, the social costs calculated relate only to the residential sector.

The procedure used to calculate these costs is:

1. The expected quantities of winter gas to be consumed by the residential sector for each of the years 1973 to 1990 for each fuel policy are calculated.
2. The price of gas per thousand therms is estimated for each future year assuming elasticities of winter supply of 3 and 5.[b] The basic price for gas used in the analysis is the price calculated by the engineering cost model, described in Appendix C, for the median building size (two dwelling units) of residential gas users in the winter of 1971. This price is $108.30 per 100 therms.

[a]From Peoples Gas, Light and Coke Company records.

[b]The elasticities of winter supply assumed are greater than unity which implies that the long-run supply curve in Figure 2–4 is concave rather than convex as drawn.

3. The difference in the price of gas between that estimated under a sulfur law and that estimated under each of the fuel ban policies is calculated for each year.
4. The price difference in each year is then multiplied by the corresponding residential winter gas consumption.
5. The present value and the annualized costs are calculated using 10 and 20 percent discount rates.

The present values and the annualized costs are presented in Table G-1. These costs are *marginal* with respect to the sulfur law.

EXOGENOUS PRICE CHANGES

Coal Ban

Now assume gas and oil prices increase independently of the coal ban policy. This may occur because of increased exploration, drilling, and transportation expenditures needed to maintain the reserve production ratio of 13 and supply the additional amount of gas demanded (the regulated well-head price of natural gas has been argued to be a major factor in the decline of the gas reserve production ratio). Regardless of the reasons, the price increase affects the costs of the coal ban in two ways: (1) fuel savings resulting from conversions from coal to oil or gas are made smaller, and (2) the rate of conversions to these fuels under the low sulfur law will decline.

In the following we consider the case in which the gas price to consumers increases by 50 percent in 1973 and stays at that level until 1990. Oil prices are assumed to increase by the same percentage, implying that the annual

Table G-1. Additional Social Costs to the Residential Sector Due to an Endogenous Gas Price Increase (Millions of Dollars)

	Present Value[a]		*Annualized Cost*	
Discount Rate	*10%*	*20%*	*10%*	*20%*
Elasticity of Gas Supply is 3				
1975 Coal Ban	5.89	4.18	.65	.72
1977 Coal Ban	4.05	2.58	.45	.45
1975 Coal and Oil Ban	35.87	23.63	3.99	4.09
1977 Coal and Oil Ban	28.90	17.68	3.21	3.06
Elasticity of Gas Supply is 5				
1975 Coal Ban	3.79	2.68	.42	.45
1977 Coal Ban	2.68	1.66	.30	.29
1975 Coal and Oil Ban	21.96	14.58	2.44	2.52
1977 Coal and Oil Ban	17.56	10.74	1.95	1.86

Note: Costs are relative to the sulfur law.
[a]Present values are for 1973.

cost differential between gas and oil increases by 50 percent. However, for simplicity we assume no change in the conversions from oil to gas and in the distribution of conversions from coal between oil and gas.

Obviously, if a coal ban is imposed by 1975 an increase in oil and gas prices would not change the end result—i.e., no coal-using dwelling units will be left by 1975. However, the estimated marginal cost of the coal ban would change. The effect of an exogenous price change on conversions is evaluated similarly to the effects of the coal price change for the low sulfur law described in Chapter 2. The procedure used to alter the sulfur law trend is depicted in Figure G-1 and described in the text that follows.

Define[c]

$$G = \frac{AB}{AC} = \frac{(C_c^{73} - C_{og}^{73}) + (C_c^{66} - C_{og}^{66})}{(C_c^{71} - C_{og}^{71}) + (C_c^{66} - C_{og}^{66})},$$

where

C_{og}^{73} ≡ the annual cost of space heating with oil and gas given the 50 percent exogenous price increase;

C_{og}^{k} ≡ the annual cost of space heating with oil and gas using the prices in years k = 1971 and 1966;

C_c^{k} ≡ the annual cost of space heating with coal using the coal prices in years k = 1973, 1971, and 1966.

Coal prices in 1973 are assumed to be equal to the 1971 price—i.e., $C_c^{73} = C_c^{71}$. Furthermore, the 1966 fuel prices are assumed to represent the fuel prices prevailing in the 1962 to 1968 period. Finally, the share of gas and oil used to calculate the C_{og}^{k} variables are 89 and 11 percent, respectively.

The ratio G is used to calculate the number of coal using dwelling units over time assuming a sulfur law with an oil and gas price increase of 50 percent. Given this revised series of coal-using dwelling units, the costs of the coal ban policy are reevaluated (conversion costs, extra labor and capital costs, and fuel savings). Recall that the cost estimations discussed in Chapter 2 did not include labor savings as part of the fuel savings of the policy. We assumed that the janitorial labor force did not have positive opportunity costs, because of

[c]Note that the following relations are implied from this formulation:

$\partial G/\partial C_c^{73} > 0; \partial G/\partial C_{og}^{73} < 0; \partial g/\partial C_c^{71} < 0;$

$\partial G/\partial C_{og}^{71} > 0; \partial G/\partial C_c^{66} < 0; \partial G/\partial C_{og}^{66} > 0.$

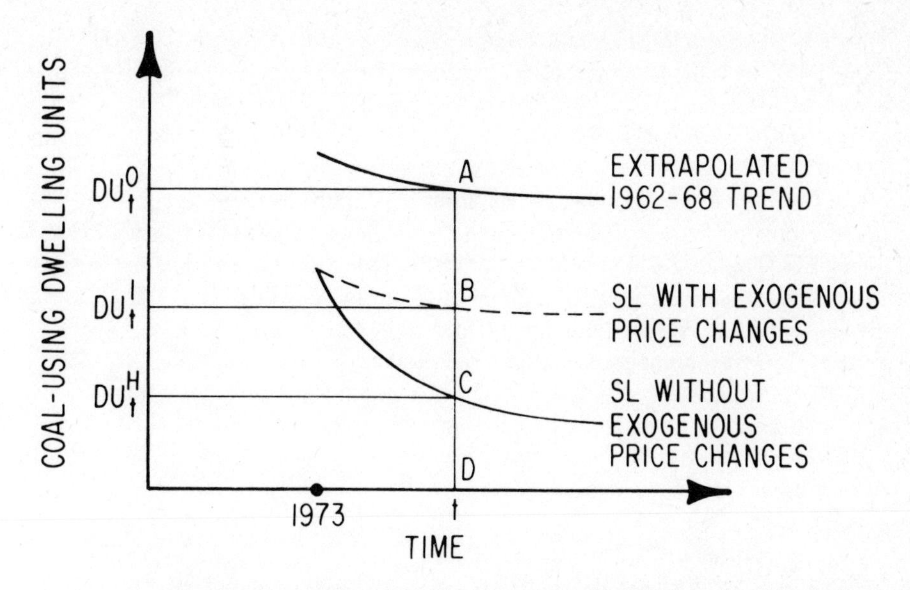

Figure G-1. Number of Coal-Using Dwelling Units with No Control and Sulfur Law.

their age distribution and skill mix. However, this assumption is not relevant when the landlord considers converting from coal. He might save some of the labor costs (some might be viewed as fixed costs). This is the reason for calculating two values of G. The first includes labor costs as an incentive for conversion; the other excludes it. The actual situation is somewhere between the two. Thus, the two extreme cases define the region relevant for the calculations. The data used for calculating the value of G are provided in Table G-2.

Table G-2. Data for Calculating the Adjustment Factor G (Dollars)

Variable	*Labor Costs Included*	*Labor Costs Excluded*
C_{og}^{73}	2793.7	2709.5
C_{c}^{71}	4875.5	3464.5
C_{og}^{66}	1630.4	1583.4
C_{c}^{66}	2597.7	1446.6
C_{og}^{71}	1870.5	1806.3
Value of G	$G_1 = .768$	$G_2 = .406$

Note: The figures correspond to a building size of 22 dwelling units.

Define DU_t^1 to be the number of coal-using dwelling units in year $t (t > 1973)$ with the exogenous price changes. Then

$$DU_t^1 = (G) DU_t^H + (1 - G)\, DU_t^0 \quad (1)$$

where DU_t^0 denotes the number of coal-using dwelling units corresponding to the extrapolated 1962 to 1968 trend and DU_t^H to the low sulfur law with 1971 prices. For simplicity G is assumed constant for the entire 1973 to 1990 period. Table G-3 contains the adjusted number of coal-using dwelling units. Using the above method, the predicted number of coal-using dwelling units for 1973 exceeds the actual number existing in 1972, when G_1 was assumed, and in 1973 and 1974 when G_2 was assumed. Therefore, the numbers were adjusted by interpolating between the 1972 actual value and the first prediction below it. Table G-4 summarizes the adjusted costs of a coal ban in 1975 and 1977 with an exogenous price increase of 50 percent for gas and oil.

Coal and Oil Ban

When the coal and oil ban is considered, an exogenous increase in oil and gas prices by 50 percent actually lowers the costs of banning oil in addition to coal because the absolute difference between annual space heating

Table G-3. Coal-Using Dwelling Units, Given a Price Increase for Gas and Oil of 50 Percent (Thousands)

Year	DU_t^0	DU_t^H	DU_t^1 *with* G_1	DU_t^1 *with* G_2
1972	207	138		
1973	192	117	131.9	135.8
1974	178	100	125.9	133.7
1975	164	84	110.5	131.5
1976	152	72	98.6	119.5
1977	140	61	87.2	107.9
1978	130	51	77.2	97.9
1979	120	43	68.6	88.7
1980	110	37	61.2	80.4
1981	103	31	54.9	73.8
1982	95	26	48.9	67.0
1983	89	22	44.2	61.8
1984	82	18	39.2	56.0
1985	76	15	35.3	51.2
1986	71	13	32.3	47.5
1987	65	11	28.9	43.1
1988	61	9	27.3	39.9
1989	56	8	23.9	36.5
1990	52	6	21.3	33.3

Note: Symbols are defined in text of this appendix.

costs with oil and gas widens. The increase of the cost differential would increase the rate of conversion from oil to gas relative to that observed under the sulfur law and also would increase the fuel saving of those dwelling units that do convert by 50 percent. There were 162 thousand oil-using dwelling units in 1972. The annualized fuel saving due to banning oil in addition to coal by 1975, with 1971 prices prevailing, was $4.08 million for a 10 percent discount rate. An increase in gas and oil prices by 50 percent increases this fuel saving by 50 percent–i.e., the annualized cost calculated for banning oil in addition to coal by 1975 should be reduced by $2.04 million (the savings are increased by this amount).

If gas prices alone are increasing by 30 percent, while oil prices stay at their 1971 level, or if any other combination of price increases takes place such that the oil-gas heating cost differential is eliminated, then the marginal cost of banning oil in addition to coal by 1975 would increase annually by about $4.0 million.

For a coal and oil ban in 1977 with a joint increase in gas and oil prices by 50 percent, the fuel saving due to banning oil in addition to coal increases by $1.65 million annually for a 10 percent discount rate. An increase in gas and oil prices (as mentioned above) that eliminates the space heating cost differential will increase the marginal costs of banning oil in addition to coal by $3.7 million annually.

Table G-4. Summary of Costs of Coal Bans in 1975 and 1977, When Gas and Oil Prices Increase by 50 Percent in 1973 (Millions of Dollars)

Adjustment Factor *Discount Rate*	G_1 *10%*	G_1 *20%*	G_2 *10%*	G_2 *20%*
I. Coal Ban 1975–Basic Assumptions.				
Coal Marketing Losses	.93	1.01	1.16	1.22
Costs of Earlier Furnace Conversions	1.20	1.81	1.54	2.83
Additional Fuel Costs	(2.21)	(2.40)	(2.76)	(2.92)
Total Costs	(.08)	.42	(.06)	1.23
II. As I, but cost of conversions are twice those presented in Table 2-2.				
Total Costs	1.12	2.23	1.48	4.06
III. Coal Ban 1977–Basic Assumptions				
Coal Marketing Losses	.73	.74	.96	.96
Costs of Earlier Furnace Conversions	.96	1.68	1.32	2.22
Additional Fuel Costs	(1.75)	(1.77)	(2.29)	(2.25)
Total Costs	(.06)	.65	(.01)	.93
IV. As III, but cost of conversions are twice those presented in Table 2-2.				
Total Costs	.90	2.33	1.31	3.15

Note: Values in parentheses are negative costs–i.e., savings. All numbers are annualized.

Appendix H

Formulas and Regression Estimates for Benefit Analysis

This appendix presents details of the benefit estimation method of Chapter 4. Separate subsections discuss in turn each of the major components of that analysis: aggregation of benefits over many locations and years, estimation of effects on the expected length of life, evaluation of extensions of longevity, evaluation of household materials benefits, and utilization of the property value approach. The last two sections discuss the relationship between the form of the marginal benefit function and the magnitudes of policy benefit estimates, and compare the "total cost of air pollution" estimate implied by this study with those presented in some other major studies of air pollution damages.

BENEFIT FORMULA WITH SPATIAL AND TEMPORAL DETAIL

This section describes the formula used to estimate the aggregate willingness to pay for air quality improvements that vary spatially and temporally. The values presented in Table 4–7 represent the social benefits of one $\mu g/m^3$ improvements in air quality that extend indefinitely into the future. Since the impacts of the residential fuel policies occur progressively over a number of years, it is necessary to estimate the social benefits of each year's improvement in air quality separately. The present value of the additional benefit of applying a more stringent policy is estimated by calculating the yearly benefits of this policy and of the original policy, taking the differences between the benefits for each year, and discounting them back to the present. Since the air quality improvements also vary geographically, this calculation must be performed separately for different locations within the region. Total regional social benefits are estimated by summing the present value estimates for each location.

The formula for calculating the social benefits of a policy in a specified geographical location and with a planning horizon of 18 years (1973 through 1990) is:

$$\sum_{i=1}^{n} \sum_{t=1}^{18} \left\{H_i \, [(B \, \Delta P_{it}^1 + C \, \Delta S_{it}^1) - (B \, \Delta P_{it}^2 + C \, \Delta S_{it}^2)] \, / (1 + r)^t \right\}$$

where

H_i ≡ number of households in location i;

B ≡ willingness to pay per $\mu g/m^3$ reduction of suspended particulates (annual arithmetic mean), per household;

C ≡ willingness to pay per $\mu g/m^3$ reduction of sulfur dioxide (annual arithmetic mean), per household;

$\Delta P_{it}^k \equiv P_{it}^k - P_{i,t-1}^k \equiv$ improvement in annual suspended particulate level from year t -1 to year t in location i;

$\Delta S_{it}^k \equiv S_{it}^k - S_{i,t-1}^k \equiv$ improvement in annual sulfur dioxide level from year t - 1 to year t in location i;

k = 1 with the policy and 2 without the policy;

r ≡ the social discount rate.

This formula is equivalent to:

$$r \sum_{i=1}^{n} \sum_{t=1}^{18} \left\{H_i [B(P_{it}^2 - P_{it}^1) + C(S_{it}^2 - S_{it}^1)] \, / (1 + r)^{t+1} \right\}$$

Since $(P_{it}^2 - P_{it}^1)$ and $(S_{it}^2 - S_{it}^1)$ are the net impacts of the policies on suspended particulate and sulfur dioxide levels in year t, respectively, this is computationally more efficient than the initial version of the formula. An analogous formula for policy evaluation using the property value method is generated by substituting the value of residential property at location i (in dollars) for H_i and letting B and C represent the willingness to pay per $\mu g/m^3$, per dollar's worth of residential property.

AIR POLLUTION AND MORTALITY RATES

The mortality analysis described in Chapter 4 was done for four demographic classes: (1) white males, (2) white females, (3) non-white males, (4) non-white females. The following describes the procedure used to estimate the effect of a one $\mu g/m^3$ reduction of particulates on the expected length of life in each of these classes for 100 age cohorts.[1]

Let the cohorts be numbered $i = 1, 2, \ldots, 100$, where i is the person's age one year after the air is improved. Let m_j be the probability of dying during the following year for a person who has just reached his $(j - 1)^{th}$ birthday. For example, m_1 equals the proportion of newborn infants who die during the first year of life and m_{25} equals the proportion of persons becoming 24 years old who die before their 25th birthday.

For cohort i, the probability of dying during the first year after an air quality improvement is $d_{i1} = m_i$; the probability of dying during the second year is $d_{i2} = (1 - m_i)m_{i+1}$; during the third year, $d_{i3} = (1 - m_i)(1 - m_{i+1})\,m_{i+2}$; etc. The expected length of life was calculated as

$$d_{i1} + 2d_{i2} + 3d_{i3} + \ldots + (102 - i)d_{i,102-i}\,.$$

(It is arbitrarily assumed for purposes of the analysis that persons reaching age 100 die within the next year; this does not affect the benefit calculations.) The mortality rates used in the analysis were 1959 to 1961 Chicago SMSA averages.[2]

Lave and Seskin estimated coefficients of regression equations of the form

$$m_i(k) = \ldots + a_i P(k) + \ldots,$$

where $m_i(k)$ is a mortality rate for demographic class i in city k, and $P(k)$ is an annual average suspended particulate figure for the city.[3] Mortality rates m_i' for Chicago assuming a one $\mu g/m^2$ pollution reduction were estimated as

$$m_i' = m_i - (m_i/m_i^L)a_i,$$

where m_i^L is the mean mortality rate for age i for the cities in the Lave-Seskin sample, m_i is the Chicago SMSA mortality rate, and a_i is the Lave-Seskin age-specific regression coefficient. This assumes that a one $\mu g/m^3$ improvement in particulate air quality affects mortality rates in Chicago, relative to other cities in the Lave and Seskin sample, in direct proportion to the ratio of the mortality rate in Chicago to the average mortality rate in their sample for each demographic class. Their reported a_i's for ages 15 to 44 and 45 to 64 were used directly. Their "all ages" coefficient was used for ages less than 15 and over 64. In their analysis all of these coefficients were significantly different from zero at the 10 percent level. To determine the impact of a one $\mu g/m^3$ reduction in particulate level, expected lifetimes are recalculated using the m_i''s instead of the m_i's. Some of these differences are reported in Table 4-1.

The mobility of population within the United States and changing air quality levels over time make regression analyses of mortality rates and applications of their results imprecise. Strictly speaking, the procedure is justified only if the members of the populations analyzed are subject to the same air quality levels throughout their lives. If, on the average, people have been subjected to generally better levels of air quality in the past than those prevailing at the time of the Lave and Seskin observations, the regression analysis overestimates the mortality rate coefficient. If the reverse occurred, then the method underestimates the coefficient. There is no basis in the present case for ascertaining which condition, if either, pertains.

EVALUATING THE EFFECTS ON DEATH AND ILLNESS RATES

Increases in Earnings

The present value of expected future earnings less expenditures for individuals was calculated using the two sets of mortality rates described above. The difference between the two present values is the contribution to income gains from a one $\mu g/m^3$ improvement in air quality.

For a person in the ith cohort, let $s'_{i1} = 1 - m'_i$ = probability of living through the first year after the policy goes into effect, $s'_{i2} = (1 - m'_i)(1 - m'_{i+1})$ = probability of living through the first two years after the air quality is improved, and so forth. Then the desired present value if the air quality does improve is

$$\begin{cases} \sum_{j=1}^{66-i} s'_{ij} E_{i+j-2}\, \delta_j + E' s'_{i,66-i}\, \delta_{66-i} & \text{if } i \leqslant 65 \\ 0 & \text{if } i \geqslant 66 \end{cases}$$

where E_k = excess of earnings over expenditures per year for persons of age k, and δ_k = discounting factor $(1.02/1.10)^k$, reflecting an assumed 2 percent real yearly growth in E and discounting future values back to the present at 10 percent. E' is the number of days' increase in expected longevity for someone who reaches age 65 times the excess of earnings over expenditures per day at age 64. Without an air quality improvement, the term involving E' is omitted and m_i's are used instead of m'_i's.

Deferral of Final Illness and Burial Expense

The savings from deferring expenses at the time of death are estimated by calculating their present value with the two sets of m_i's and then computing the difference. The expected present value of these expenses for someone in the ith cohort is

$$C \sum_{j=1}^{102-i} d_{ij}\,\delta_j,$$

where C = burial and final medical expenses (assumed to be \$2000), and the other symbols retain their previous meanings.

Reduction in Health Care Expenses

Estimates of health care expenses were made by computing the present value of the average net reduction in health care costs over an individual's future lifetime. The reduction is assumed to be H = \$.43 per year, if particulates or SO_2 are reduced by one $\mu g/m^3$. The formula is

$$H \sum_{j=1}^{102-i} s_{ij}\,\delta_j$$

for someone in the ith cohort.

Expenditures Diverted from Other Uses

The total present value of future expenditures is

$$X_i' = X_i \sum_{j=1}^{102-i} s_{ij}\,\delta_j,$$

where X_i = current expenditure level per person for the demographic group. The income diverted from other uses was then calculated as $(\Delta L_i/L_i)X_i'$ where ΔL_i = change in expected length of life, and L_i represents the expected length of life.

MATERIALS MAINTENANCE BENEFITS

The present value of a series of constant expenditures of \$$V$, the first immediately, the second after a time duration of T, and the others periodically at intervals of length T thereafter is

$$V \sum_{i=0}^{\infty} e^{-riT} = V/(1 - e^{-rT}).$$

Salmon estimates that exterior surfaces are repainted every six years in rural areas and every three years in urban areas.[4] Using these figures with data on average air quality levels in rural and urban areas in the United States, it was calculated that a one $\mu g/m^3$ improvement in SO_2 within an urban area might reduce painting frequencies by an average of 8.5 days. The present value defined above was evaluated with $T = 4$ years corresponding to the pre-improvement case and again with $T = 4$ years plus 8.5 days for the post-improvement case. An estimate of $V = \$125$ was based on a study of surface areas of houses in Chicago assuming a 20¢ per square foot painting cost (from Salmon), and assuming further that one-fourth of the residences in Chicago have exterior painted walls instead of brick, glass or other materials. The resulting estimate of gain per household was $1.79. Similar procedures were followed for the other materials. Table H-1 reports the principal parameters in this work.

INFERENCE FROM PROPERTY VALUE STUDIES

Crocker used statistical regression techniques to estimate the constants (*a, b,* etc.) in the equation

$$PR = a\, P^b\, S^c\, X_1^{d_1}\, X_2^{d_2} \ldots X_n^{d_n}\, u \qquad (1)$$

where

$PR \equiv$ sale price plus incidental costs of purchase of FHA-insured dwellings in the sample;

$P \equiv$ annual average concentration of suspended particulates for the year preceding the year of sale;

$S \equiv$ annual average concentration of sulfur dioxide for the year preceding the year of sale;

$X_1, \ldots, X_n \equiv$ other variables, including income level of the neighborhood (from census tract data), school quality in the neighborhood, distance from downtown Chicago and from Lake Michigan, size of house, size of lot, term of mortgage in years, and indicators of various structural characteristics of the dwellings;

$u \equiv$ the "disturbance," a random variable encompassing the combined effects of all the variables not specified explicitly.[5]

Table H-2 contains Crocker's estimates of the constants b and c for three variants of the regression equation. Variant 1 is the equation as described above; variant 2 excludes the distance from downtown Chicago and Lake Michigan from among the X_i variables; variant 3 is the same as variant 1 but includes also, as independent variables, indicators of the monthly variation in suspended particulate and sulfur dioxide levels around the yearly averages.[6]

Table H–1. Assumptions and Results of Materials Maintenance Analysis

Maintenance Operation	*Initial Frequency in Urban Areas*	*Cost of One Maintenance Operation per Household (Dollars)*	*Pollutant Affecting Frequency*	*Decrease in Frequency from 1 μg/m³ Reduction in Pollutant (Days)*	*Saving in Present Value of Maintenance Cost per Household (Dollars)*
Exterior Painting of Residences	4 years	125.00	SO_2	8.5	1.79
Interior Painting of Residences	5 years	900.00	SO_2	4.3	6.18
Laundry, Frequent Cleaning of Interior Surfaces	1 week	3.00	Particulates	0.07	15.30
Dry Cleaning and Other Infrequent Cleaning	6 months	13.50	Particulates	1.8	2.67
Window Cleaning	6 months	24.00	Particulates	1.8	4.75

Note: The figures in the last column were doubled when carried to Table 4–4, to account for increased value of materials in use, between maintenance operations.

Table H-2. Regression Coefficients and Derived Estimates of Willingness to Pay for Air Quality Improvements

		Coefficients of:		Willingness to pay for 1 $\mu g/m^3$ reduction in:	
Year	Equation Variant	Particulates (b)	SO_2 (c)	Particulates	SO_2
1967	1	-.3537 (.1051)	.0296 (.0593)	$2.55	~0.0
1967	2	-.3625 (.1057)	-.0828 (.0441)	2.62	$1.63
1967	3	-.4453 (.1133)	-.1784 (.1525)	3.21	3.52
1965	1	-.2207 (.1616)	-.1304 (.0270)	1.72	3.97
1965	2	-.3560 (.1097)	.0306 (.0192)	2.78	~0.0
1965	3	-.3163 (.1693)	-.1521 (.0351)	2.47	4.63

Note: Standard errors are given in parentheses below each coefficient estimate. Variant 2 excludes distance variables; variant 3 includes the variance and skewness of the pollution variables. Willingness to pay means the estimated increase in value of $1000 of residential property if the level of the pollutant is reduced by one $\mu g/m^3$, *cet. par.* Two estimates of the SO_2 coefficient (c) are positive: the .0296 is not significantly different from zero at the 10 percent level; the .0306 is, but it is rejected on grounds of incompatibility with the knowledge that SO_2 has adverse rather than beneficial effects. The other coefficients are all significantly different from zero at the 10 percent level.

Source: T.D. Crocker, *Urban Air Pollution Damage Functions: Theory and Measurement*, National Technical Information Service Publication No. PB–197665 (Springfield, Va., 1970), pp. 55, 57, 74.

The table also includes estimates of the difference in property value in dollars, per $\mu g/m^3$ change in S and in P, per \$1000 of property value. The differences in the average pollution levels of Crocker's two samples (1965 and 1967) were used to estimate these figures. Equation 1 implies that when pollution levels vary from location to location, the difference in property value would be

$$\Delta PR = PR' \, (b \, \Delta P/P' + c \, \Delta S/S') \qquad (2)$$

where

ΔPR ≡ approximate estimated difference between locations in the value of residential property;

PR' ≡ average value of PR (= \$16,208 in Crocker's 1965 data and \$17,154 in the 1967 data);

ΔP and ΔS ≡ differences in pollutant levels between locations;

P' and S' ≡ average pollution levels within Crocker's samples ($P' = 138.54$ and $S' = 50.7$ in Crocker's 1967 sample and $P' = 128.05$ and $S' = 32.9$ in his 1965 sample).[7]

Thus, $b(PR'/P')$ is an estimate of a property owner's willingness to pay for a one $\mu g/m^3$ reduction in the level of P, if his property is valued at PR' and the initial particulate level is P'. Similarly, $c(PR'/S')$ estimates the willingness to pay for a one $\mu g/m^3$ reduction in S. The estimates of the willingness to pay in Table H-2 are values of (\$1000) (b/P') and (\$1000) (c/S').

Using Crocker's regression coefficients in this way assumes they are appropriate for a wider range of property values and initial levels of pollution than those in his sample. In this analysis, estimates of willingness to pay using Equation 2 are based on the slope of Equation 1 at the mean (as illustrated by the dashed line in Figure H-1). This is the simplest possible way to extrapolate beyond the range of Crocker's data. Moreover, it avoids generating unreasonably inflated estimates of benefits, as would be the case if the extrapolation were based on Equation 1 itself, which is of log-linear form (illustrated by the solid line in Figure H-1).

A second method of estimating willingness to pay using studies of rates of return to productive resources at different locations is to examine wages in different cities with various pollution levels. Izraeli has done this in a regression analysis framework structurally similar to the property value work.[8] He estimates that wages are raised 0.06 percent for every 1 percent increase in sulfates (a chemical constituent of suspended particulates), adjusting for the influence of other variables. Thus, a worker earning \$10,000 per year in a city having a 1 percent higher level of sulfates than another city would be willing to give up \$6.00 per year in wages if he relocated to the other city when the effects

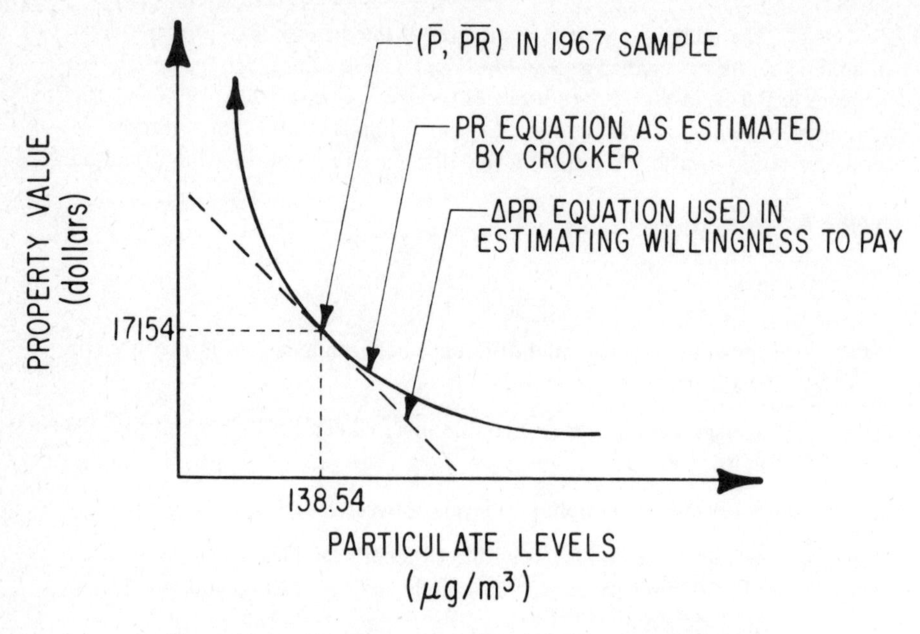

Figure H–1. Regression Equation and Estimating Equation.

of air pollution alone are considered. To compare this with estimates made using the property value method, recognize that a difference of \$5.00 per thousand dollars of residential property value (as would be associated with a 1 percent change in pollution levels) is equivalent to a change of \$100 in the value of a \$20,000 house. This is approximately \$11 annually, or about 1.5 times the estimate of annual willingness to pay made using the wage analysis.

THE FORM OF THE BENEFIT FUNCTION

The definition of the federal secondary standards suggests that there is no additional benefit from improving air quality beyond the standards. If this is true, then it is reasonable to postulate further that the additional gains from air quality improvement diminish as this level is approached. Figure H–2 illustrates one such a relationship.

This issue poses a dilemma for anyone who wishes to quantify the benefits of a single component of an overall air quality improvement program. The present study is a case in point. The Illinois implementation plan embodies restrictions not only on residential fuel use, but also on power plants industry, commercial establishments, and automobiles; residential fuel policies contribute only a small proportion of the total air quality improvement sought for the Chicago area. The problem is: Should benefit computations be based on a valua-

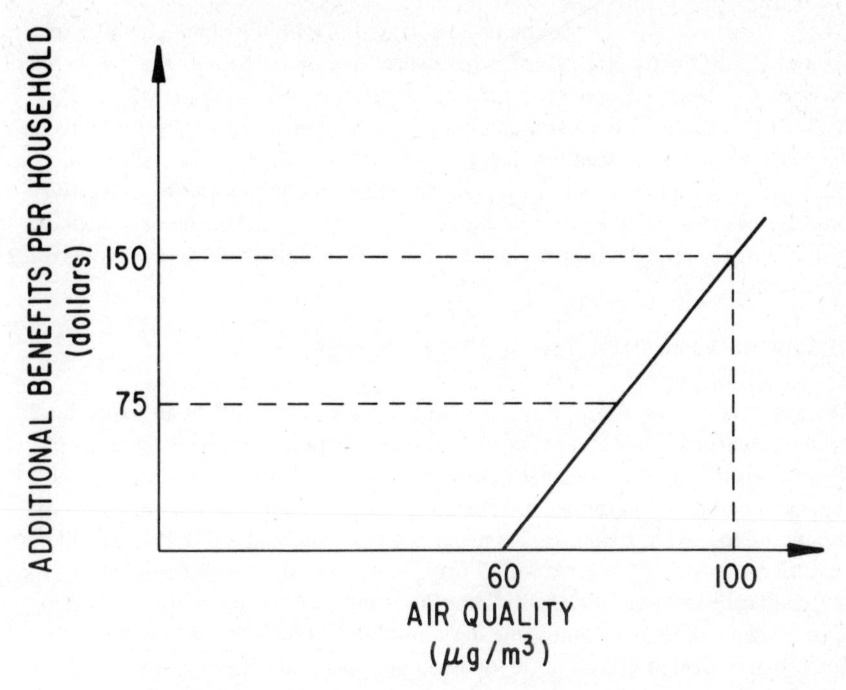

Figure H-2. Additional Benefits from One $\mu g/m^3$ Improvement in Air Quality.

tion which corresponds to the pollution levels in 1968 before the imposition of stringent controls, to the level of improvement sought by the control plan, or a level between the two? In other words, should the policy be evaluated as though it is the first or the last, among all emission controls being introduced?

To answer this question, the policies should not be considered separately from one another. Benefits and costs for the proposed controls on each economic sector should be evaluated, as is done for residences in this study. The ratio of benefits to costs should be computed for each policy assuming each one is to be the first. A ranking should be made, based initially on the benefit-cost ratios, but possibly modified in the light of administrative and legal considerations, so that the ranking reflects as well as possible a public order of preference between the policies. This ranking then defines the proper order of policies for benefit evaluation. The benefits should then be recalculated. They will be reduced for all but the first policy in the list. The optimal decision is to select policies beginning from the top of the list, accepting only the policies for which recalculated social benefits exceed their social costs.

Carrying through this plan expands the scope of analysis required to evaluate any policy several fold in comparison with this study. An alternative approach would be to use benefit valuations slightly lower than those inferred from damage studies—that is, the benefit valuation would be based on a figure such as \$75 in Figure H-2. This is what was done in this chapter: the maximum estimate was purposely rejected; instead, an estimate was accepted which in all likelihood is based on market indications of under-perceived values of pollution damages. What is sacrificed is a degree of precision of the benefit evaluation, since the valuation used is arbitrary. Furthermore, a comprehensive air quality program based on such a procedure may tend to over-control, since the benefits of the least favorable among the policies being considered are overestimated using this method.

COMPARISON WITH OTHER BENEFIT STUDIES

The above reasoning is important in estimating the "total cost of air pollution," or in other words, the total benefit of reducing air pollution from, for example, 1968 levels in major urban areas before stringent regulations were initiated, to the levels embodied in the federal standards. Suppose that the marginal benefit function depicted in Figure H-2 is linear between the points (60,0) and (100,150). Its formula is then $B^* = 3.75x - 225$ for $x \geqslant 60$, and 0 for $x < 60$, where x represents the level of "air quality" and B^* is the additional benefit per household.[a] Then the total present value per household of benefits of improving air quality from 100 to 60 is \$3,000, or \$300 per year, annualizing with a 10 percent discount rate. Table 4-6 suggests that about \$150 of this represents benefits in the material damage area, and about \$150 is related to health effects.

A comparison of these estimates with those obtained by other investigators may be of interest. This will be done in relation to four comprehensive studies of benefits of air quality improvement, by O'Connor, the Committee on Air Pollution, Ridker, and Barrett and Waddell.[9] These studies vary in purposes and approach. The first two were written in response to essentially local air quality problems: smoke from the burning of coal in Pittsburgh in the early part of this century, and the great London pollution episode of 1952. The last two are more academic in nature and present estimates of the aggregate cost of air pollution for the United States as a whole. The first and second are the products of subcommittees working within large but localized groups studying many aspects of air pollution damages and abatement. The third is an essentially individual effort; the fourth is a synthesis of numerous other pieces of recent literature having a bearing on the subject.

Translating the estimates into per household terms, Barrett and Waddell's is very close to the \$300 calculated in the present study; theirs is \$152

[a]The federal secondary standards for both particulates and sulfur dioxide are 60 $\mu g/m^3$, which implies a zero benefit for additional air quality improvements. Annual average particulate and sulfur dioxide levels in much of Chicago in 1968 were 100 $\mu g/m^3$. Table 4-6 indicates a maximum benefit of 150 dollars for a one $\mu g/m^3$ simultaneous improvement in both particulates and sulfur dioxide.

per household for health and $72 for materials benefits. They also present an estimate equivalent to $150 per household based on the property value method. Their health estimate is based on Lave and Seskin's regressions, as is the one presented here. Their estimate of materials benefits is based largely on work by Salmon,[10] plus a few other recent studies of damages to specific kinds of materials. It represents primarily corrosion damage—that is, it covers repainting plus damages to other materials (mainly outside of the residential sector) not considered in this study. Also, their estimate does not include soiling losses, which are included in the estimates of this chapter. Their property value estimate is based on several recent studies other than Crocker's,[11] all of which imply a damage coefficient of the same order of magnitude. It is of interest that many recent property value studies imply a valuation of air quality improvement at about the same dollar value.

Ridker's results are substantially different from those presented in this study and those presented by Barrett and Waddell. Ridker's estimate of $360 to $400 million annual health costs for the United States [12] is about $10 per household (assuming that these costs are borne by the 40 million households in the urban areas of the United States.) This is a full order of magnitude less than the other estimates. His assessment of potential gains from reduced soiling is based on a survey of households in Philadelphia and on interviews with industrial firms. The survey results are generally negative in the sense that household cleaning frequencies in different neighborhoods do not appear related to air pollution. This observation is not necessarily inconsistent with an assignment of positive dollar costs to soiling caused by air pollution. It may be that individuals choose to bear the cost in the form of dirtier household surfaces rather than increasing expenditure of time and money for cleaning, as pollution increases. A more detailed, recent survey study by Booz-Allen and Hamilton, Inc., and National Analysts, Inc., does exhibit significant correlation between pollution and cleaning frequencies for some types of household surfaces.[13]

The other two studies are of interest more as checks against the comprehensiveness of the present investigation than for their numerical estimates, since one is 60 years old and the other was conducted in a foreign country and is 20 years old. Neither of these investigations placed a comprehensive dollar value on adverse health effects of air pollution; the analytical difficulties of this at the time were considered insurmountable by both groups. O'Connor's estimate of the soiling cost of Pittsburgh air pollution is about $35 per household per year (in 1913 dollars) for categories roughly comparable to those considered in the materials section of the present work, and $50 per household for others, such as damage to sheet metal work and nonresidential buildings, and increased artificial lighting necessitated by air pollution. The British study estimates $43 per household for material damages roughly corresponding to those considered here, and nearly $100 for reduced efficiency in provision of transport services, corrosion of metals, damage to non-residential structures, and damage to stocks in retail stores. (Their figures in British pounds have been converted into 1952 dollars.) On the basis of these studies, one might be led to double the $150 material damage estimate of the present study and attribute the rest to damages outside the residential sector. (These other estimates are used solely for the purpose of estimating the ratio between values of damages considered in this chapter and other damages; the dollar estimates for Pittsburgh in 1913 and London in 1952 are clearly out-of-date.)

On the whole, however, the order of magnitude of the estimates presented in this chapter appears substantiated by other major existing studies of air pollution damages. The estimate is about $300 annually, per household, as the total "cost of air pollution" (in the sense of foregone earnings, as well as out-of-pocket expenditures associated with pollution damage), which could be avoided by reducing air pollution levels from prevailing levels in urban areas in the mid-60s to federal standards levels. Restrictions on residential coal burning (the sulfur law and coal ban) have been shown to eliminate, on the average, about 10 percent of this total cost in Chicago.

NOTES TO APPENDIX H

1. The approach taken in this appendix for estimating expected lifetimes and present values of future earnings is based on a paper by R.A. Schrimper, "Estimating Health Costs," Mimeographed (Chicago: University of Chicago, 1973).
2. E.A. Duffy and R.E. Carroll, *U.S. Metropolitan Mortality, 1959–61,* U.S. Public Health Service Publication No. 999–AP–39 (Cincinnati: U.S. National Center for Air Pollution Control, 1967), Table 5, p. 6.
3. L.B. Lave and E.B. Seskin, "Does Air Pollution Shorten Lives?" in *Proceedings of the Second Research Conference of the Inter-University Committee on Urban Economics,* pp. 293–328.
4. R.L. Salmon, *Systems Analysis of the Effects of Air Pollution on Materials,* National Technical Information Service Publication No. PB–209192 (Springfield, Va., 1972).
5. T.D. Crocker, *Urban Air Pollution Damage Functions: Theory and Measurement,* National Technical Information Service Publication No. PB–197668 (Springfield, Va., 1970).
6. *Ibid.,* pp. 55, 57, 74.
7. *Ibid.,* pp. 51, 52.
8. O. Izraeli, "Differentials in Nominal Incomes and Prices Between Cities," Ph.D. dissertation (University of Chicago, 1973).
9. See, for example, J.J. O'Connor, Jr., *The Economic Cost of the Smoke Nuisance to Pittsburgh,* Mellon Institute of Industrial Research Smoke Investigation Bulletin No. 4 (Pittsburgh: University of Pittsburgh, 1913); Great Britain, *Parliamentary Papers,* Vol. 8 (1953–54) (*Reports from Commissioners, Inspectors, and Others,* vol. 1), "Report of the Committee on Air Pollution," Cmd. 9322, 1954;

R.B. Ridker, *Economic Costs of Air Pollution: Studies in Measurement* (New York: Praeger Publishers, 1967); and L.B. Barrett and T.E. Waddell, *Cost of Air Pollution Damage: A Status Report,* U.S. Environmental Protection Agency Publication No. AP-85 (Research Triangle Park, N.C., 1973).

10. Salmon, *op. cit.*
11. Crocker, *op. cit.,* p. 23.
12. Ridker, *op. cit.,* p. 56.
13. Booz-Allen and Hamilton, Inc., and National Analysts, Inc., *Study to Determine Residential Soiling Costs of Particulate Air Pollution,* National Technical Information Service Publication No. PB–205807 (Springfield, Va., 1970).

About the Authors

Alan S. Cohen received his Ph.D. and M.S. degrees in Operations Research from Northwestern University and his B.S. in Industrial Engineering from the University of Massachusetts. He is presently an Environmental Systems Engineer in the Energy and Environmental Systems Division at Argonne National Laboratory, Argonne, Illinois. Dr. Cohn has participated in many environmental planning programs over the past five years and has several publications dealing with air pollution control.

Gideon Fishelson received his Ph.D. in Economics from North Carolina State University at Raleigh and his M.S. and B.S. degrees in Agricultural Economics and Agronomy from the Hebrew University, Jerusalem. Presently he is a Research Associate at the Economics Department, the University of Chicago. Dr. Fishelson published extensively on topics related to human capital, natural resources, public policy and the environment and other economic issues.

John L. Gardner is Research Associate in the Center for Urban Studies at the University of Chicago. He received an A.B. degree from Oberlin College and a Ph.D. in economics from the University of Minnesota. He has authored two articles on the benefits of air quality improvements and is engaged in research on spatial implications of environmental policies. Other areas of interest include local government finance and the economics of regional growth and inter-regional migration.